Applied equine nutrition and training

Applied equine nutrition and training

Equine NUtrition COnference (ENUCO) 2007

edited by: Arno Lindner

Arbeitsgruppe Pferd

Wageningen Academic
P u b l i s h e r s

Photos cover: Sabine Heueveldop

ISBN: 978-90-8686-040-1

First published, 2007

**Wageningen Academic Publishers
The Netherlands, 2007**

Table of contents

Articles

What's new in equine nutrition (2005-06)?   15
Patricia A. Harris

Evaluating recent equine research (2005-06)   35
Brian D. Nielsen

Noteworthy changes to the horse nutrition requirements
by the National Research Council of USA   49
Brian D. Nielsen

Relevance and standardisation of the terms Glycaemic
index and Glycaemic response   57
Patricia A. Harris and Ray J. Geor

The role of nutrition in colic   79
Andy E. Durham

Recent papers on equine nutrition related to medicine
(2005-06)   95
Simon R. Bailey

Diagnosis and management of insulin resistance and
Equine Metabolic Syndrome (EMS) in horses   107
Nicholas Frank

The application of *Vitex agnus castus* and other
medicinal herbs for the symptomatic relief of
hyperadrenocoticism and Equine Metabolic Syndrome   127
Hilary Self

Food-allergy in horses   155
Regina Wagner, Derek C. Knottenbelt and Birgit Hunsinger

What's new in equine sports nutrition (2005-06)?   161
Raymond J. Geor

News on equine sports science (2005-06)　　　　173
Arno Lindner

Expanded abstracts

The influence of low versus high fibre haylage diets in
combination with training or pasture rest on equine
gastric ulceration syndrome (EGUS)　　　　193
Andrea D. Ellis, Maarten Boswinkel and
Marianne M. Sloet van Oldruitenborgh-Oosterbaan

An investigation into the efficacy of a commercially
available gastric supplement for the treatment and
prevention of Equine Gastric Ulcer Syndrome (EGUS)　　　199
E. Hatton, C.E. Hale and A.J. Hemmings

Changes in plasma metabolites concentrations and LDH
isoenzyme activities in Thoroughbred racehorses after
racing　　　　211
Akihiro Mori, Hiroyuki Tazaki, Nobuko Mori, Kieko Tan,
Yukino Sakamoto, Masaru Hosoya, Ryuma Nuruki and
Toshiro Arai

Changes in enzyme activities in peripheral leukocytes of
Thoroughbred racehorses after racing　　　　215
Toshiro Arai, Akihiro Mori, Kieko Tan, Nobuko Mori,
Yukino Sakamoto, Masaru Hosoya and Ryuma Nuruki

Forage conservation methods - impact on forage
composition and the equine hindgut　　　　219
Cecilia E. Müller and Peter Udén

The application of mathematical models to assess the
effect of enzyme treatment or sugar beet pulp on the
rate of passage of ensiled lucerne in equids　　　227
Jo-Anne MD Murray, Ruth Sanderson, Annette Longland,
Meriel Moore-Colyer, Peter M. Hastie and Catherine Dunnett

Growth and glucose/insulin responses of draft cross weanlings fed Total Mixed ration cubes versus hay/concentrate rations   233
Sarah Ralston, Harlan Anderson and Roy Johnson

Energy intakes of three different equine populations in comparison to NRC recommendations   239
K. Wilkinson, J. Holley, C. Preston and R.J. Williams

Editorial

For the second time the Equine NUtrition COnference – Practice (ENUCO) has become reality. Many obstacles had to be cleared. However, the willingness of the invited speakers to share their knowledge with those needing the information in equine practice and my interest in establishing in Europe a meeting where this exchange can succeed on an international basis made it possible. Of course, it helps a lot that I like to travel and get to know persons in all regions on Earth. And Vienna is a wonderful place to visit as well as to work in.

This ENUCO – Practice already incorporates another subject that is of key interest to me: Training and conditioning of sport horses. The next meeting in 2009 will be an in depth platform for both subjects: Nutrition and Training. The equity is reflected in the title of this book already. I look very much forward to it!

I wish you a benefit from the content of this book!

Arno Lindner

Articles

What's new in equine nutrition (2005-06)?

Patricia A. Harris
Equine Studies Group, WALTHAM Centre for Pet Nutrition, Melton Mowbray, Leics, LE14 4RT, United Kingdom

Over the last few years there has been an increase in the number of papers published which relate to nutrition and its effects on the health, welfare, behaviour and performance of the horse. The problem comes in using this information and perhaps in determining which bits of which papers are suitable for use. The following four papers illustrate some of the work that has been carried out in two areas of interest of mine – electrolytes and laminitis. They remind us of the importance of the basic fundamentals of nutrition and highlight some of the exciting ways that nutrition can assist with the detection and management of clinical disease; they explore some of the frustrations around answering the most simple questions – how much salt should I feed my horse and how we can unlock the potential of herbs and other nutraeuticals for the benefit of the horse?

Lutersson, N., Chunekamrai, S., Estepa, J.C. and Aguilera-Tejero, E., 2005. Secondary nutritional hyperparathroidism in ponies in northern Thailand. Proceedings Equine Nutrition Conference Hannover. Pferdeheilkunde 21: 97–98.

Objectives

'To improve the clinical situation through inexpensive effort and increase the understanding of nutrition and health care among the horse owners'.

Why chosen

Today in the developed world, with relatively easy access to good commercial feeds and feedstuffs, we tend to concentrate on the role that nutrition plays in optimising performance, supporting good health and positive behaviours. We tend not to be concerned with overtly inappropriate and inadequate nutrition. This paper is therefore a welcome reminder that poor nutrition can really cause severe clinical problems.

Background

Horses fed diets with markedly imbalanced ratios of calcium (too little) to phosphorus (too much) can develop secondary nutritional parathyroidism (also known as 'big head' disease or Millers disease: NHPT). Classically this is caused by the excessive inclusion of bran into the diet without additional calcium to balance the Ca: P ratio. Clinical signs include increased size of the nose and lower jaw due to demineralisation of the bones with replacement by fibrous tissue. Additional signs can include respiratory noise, lameness, bone and joint tenderness, loose teeth and emaciation.

Overview of the study

149 ponies from Northern Thailand participated. They were initially evaluated for clinical signs of NHPT and information with respect to their feeding and management was collected. Serum samples were taken to measure Parathyroid hormone, Calcium and Phosphorus. Based on the initial data, workshops were organised where the importance of calcium and the consequences of malnutrition were discussed. The horse owners were then offered limestone at a reduced price but it was their choice whether they took up this offer or not. The ponies were then evaluated approximately every 3 months for a year.

Main findings

Many of the ponies were fed large amounts of a cheap local feed source: rice bran together with locally grown grasses, which were high in oxalates (Setaria species). At the start more than 60% of the ponies had clinical signs of NHPT and less than 10% of the owners fed additional calcium. The typical Ca: P ratio in the diet was around 0.15 –0.3:1. After just over a year more than 90% of the horse owners were using calcium supplements (with the aim of achieving a Ca: P ratio of around 1- 1.5:1) and less than 30% of the ponies had clinical signs, plus the severity of the signs had reduced.

Practical interest

This highlights the real importance of nutrition and the need for education especially where there is a lack of core understanding. This may be important even in the developing countries as more people start to own horses without any background history of horse keeping. It reminds us that we should not ignore the basics of good horse nutrition, that both intake quantity, source and balance is important and that we should not take for granted even the simplest of nutrients such as Calcium. Calcium was in fact one of the first specifically identified nutrients to have a recognised role in health, as by the early 19th century it was appreciated that calcium and phosphorus were needed for 'hard bone'.

Comments on the study

Under the circumstances the authors should be congratulated for undertaking this work. Increased numbers would have helped the data with respect to PTH and given that the last samples were taken in Nov 2004 it would be good to have an update as to the level of continued supplementation.

Some comments re Calcium

- *Principle natural sources*: leafy forages, particularly legumes; supplementation may be achieved for example with limestone flour (calcium carbonate), calcium gluconate or dicalcium phosphate. Response to supplementation may be individually variable. Tends to be bound in plant tissue with protein and organic acid anions. Calcium is a nutrient that is commonly deficient in natural diets and the addition of high phosphorus containing feedstuffs can exacerbate the situation.
- *Digestion*: Most of the calcium compounds ingested (apart from oxalates) are converted by the gastric juices to CaCl, which is almost completely dissociated into ions. Ionic calcium is the principle form that is absorbed from the duodenum (and possibly the stomach). There is apparently active transport of calcium as

well as passive or facilitated diffusion. The possible role of the Vitamin D responsive calcium binding protein in the horse is not clear although it has been identified in the duodenum of the horse. Calcium absorption had been said to decrease with age. It is also reduced with high intakes of oxalic acid (found especially in some grasses) and phytate (found in cereals). >0.5% oxalic acid in the feed may reduce calcium absorption (when the calcium: oxalate ratio is <0.5 [weight to weight basis] but with a ratio ≥ 1.5, higher oxalic acid levels [up to 0.87%] may be tolerated). High phosphorus intake (especially in form of phytate) may disturb calcium absorption, especially when the Ca:P ratio is ≤ 1.

- *Homeostasis*: A number of hormones affect calcium status, principally parathyroid, calcitonin and vitamin D but also the adrenal corticosteroids, oestrogens, thyroid hormones and glucagon. PTH principally acts when blood calcium levels fall in order to increase blood calcium concentrations and at the same time decrease the Phosphorus levels. PTH acts via effects on the urinary excretion of P and Ca as well as the rate of skeletal remodelling and bone resorption. Calcitonin interacts primarily on bone and kidney and to a lesser extent the intestine. Calcitonin acts to lower blood calcium concentration. Vitamin D's role in the horse is not extremely clear but it acts, at least in other species to increase Calcium and Phosphorus absorption from the intestine. In addition, it also has an effect on the bone where small amounts are believed to be necessary to allow the osteolytic cells to respond to PTH. The renal excretion of calcium has been reported to be related to the amount of absorbed calcium, the phosphorus level of the diet and the anion-cation balance of the diet. The major route of excretion for calcium is via the kidney.
- Guide to maintenance requirements:
  — A ratio of Calcium: Phosphorus of between 1:1 and 2:1.
  — Intakes based on the following of a minimum of 40mg/kg Bwt. (NRC, 2007)

Assumed endogenous losses	20 (mg/kg bw/day)
Assumed availability (%)	50
Minimum requirements for maintenance	40 (mg/kg bw/day)[1]

[1]the new NRC (2007) recommends. However, the author recommends a more optimal level of around 75mg/kgBW/day.

What next

How can the equine nutrition community work to provide advice and support to other regions in the world where horses are clinically affected due to poor education and the lack of simple solutions to crucial nutritionally related problems?

Reference

NRC (2007) National Research Council Nutrient Requirements of Horses, 6th edn., National Academy Press, Washington DC.

Sampieri, F., Schott, H.C., Hinchcliff, K.W., Geor, R.J. and Jose-Cunilleras, E., 2006. Effects of oral electrolyte supplementation on endurance horses competing in 80 km rides. Equine Vet J suppl 36: 19-26.

Objectives

'To ascertain whether oral administration of a high dose of sodium chloride (NaCl) and potassium chloride (KCl) to endurance horses would differentially increase water intake, attenuate bodyweight (bwt) loss and improve performance when compared to a low dose.'

Why chosen

There is currently considerable confusion around the optimal amounts and types of electrolytes that should be given to endurance horses. Using a straightforward factorial approach to replenishment seems to overestimate the daily requirements of horses, especially those that are sweating considerably. In part this may be because the content of the gastrointestinal tract provides an important reservoir for sodium during hard work and therefore the electrolyte losses that occur with sweating may not need to be restored all at once; plus the electrolyte losses in the sweat in some endurance animals may not be as high as estimated for a number of reasons. This paper provides some insight into this confusing area.

Patricia A. Harris

Background

Endurance riding is one of the fastest growing equestrian sports in many developed countries. Endurance horses are asked to undertake long duration exercise (up to 160km/day) usually at relatively low intensities but recently, at international competitions at least, the racing speeds have increased considerably (up to an average of 22.5km/hr). The ability to do this is highly dependent on body stores of fuel in the form of both glycogen and fat as well as adequate water and electrolyte provision. The evaporation of sweat is one of the major mechanisms for the removal of excess heat produced during energy utilisation. The onset of substrate depletion, hyperthermia, and disturbances to fluid, electrolyte and acid-base homeostasis may result in elimination of the horse from the ride (so called 'metabolic failure'). Bodyweight losses of 3 – 7% commonly occur during endurance rides. Anecdotal reports from elite level endurance races have indicated that approximately 30-35% of horses are eliminated before completion, with 60% eliminated due to lameness and 25% due to metabolic problems associated with severe dehydration, heat stress, synchronous diaphragmatic flutter (SDF), muscle cramping, and/or rhabdomyolysis.

Sweat production seems to only decrease after extreme water loss and although there may be some changes in sweat composition with time, basically sweat production is accompanied by an obligate loss of electrolytes. Sweat contains relatively low levels of calcium ($\sim$0.12 g/L), magnesium ($\sim$0.05 g/L) and phosphate ($<$0.01 g/L) but relatively high levels of sodium ($\sim$3.1 g/L), potassium ($\sim$1.6 g/L) and chloride ($\sim$5.3 g/L). The sodium requirements for a horse at rest have been estimated at 20mg/kg BW/day (assuming that the sodium sources are 90% available). On a factorial basis it has been suggested that the sodium requirements for exercise should take into consideration the sodium content of sweat and the amount needed to be fed to replace this ($\sim$3.45 g/L for replacement) and the amount of sweat produced i.e. a rough guide for light, moderate, hard and very heavy exercise around 0.5-1, 1-2, 2-5 and 7-8 L/100kg bodyweight. Horses under more extreme environmental conditions may sweat up to 10 – 15L/hr. *However, most nutritionists do not recommend to feed during a race sufficient salt to replenish all that is theoretically lost (due to the gut acting as a reservoir during the ride, potentially different electrolyte sweat contents, different availabilities etc.) which adds to the confusion in this area.*

Techniques for water and electrolyte replacement are, however, vitally important for helping to reduce the risk of thermoregulatory failure and other metabolic problems associated with dehydration and electrolyte disturbances. However, currently a wide variety of strategies are used with no consensus on the optimal method. Options include the administration of hypertonic electrolyte pastes before and during the race, the addition of electrolytes to meals offered at rest stops, and provision of hypotonic electrolyte solutions for voluntary consumption. The optimal level of supplementation is also not known and likely varies between horses. In practice, riders will establish the most appropriate replacement strategy for their horse by a process of trial and error, ideally through the evaluation of different approaches during training rides.

Overview of the study

A randomised, blinded, crossover study was carried out using 8 non-elite horses (6 Arabians, one QH and one mustang: mean Bwt 441.5kg: mean age 11.8 yrs) participating in two 80km rides over the same course, under similar weather conditions, but 28days apart. The horses were requested not to have been given any additional electrolyte supplementation within 48hrs of each ride. Thirty minutes before, and at 40km of the first ride, half of the horses (i.e. 4) received orally as a slurry with apple sauce 0.2g NaCl/kg Bwt (i.e. 90g for a 450kg horse) and 0.07gKCl/kg bwt (i.e. 31.5g for a 450kg horse). The other four horses received 0.07g NaCl/kg Bwt (i.e. 31.5g for a 450kg horse) and 0.02g KCl/kg bwt (i.e. 9g KCl for a 450kg horse). Total intakes were estimated by the authors to replenish the amount of Na and K lost in 30 L of sweat (high dose) or 10L of sweat (low dose) *NB using a low estimate for sodium sweat replacement (~2.5g/l).*

Main findings

All 8 horses finished the rides although in the first ride one was eliminated, and in the second, two were eliminated; all due to lameness. The water intake was estimated to be significantly greater with the higher salt intake both at the 40km mark and overall. But there were no differences in estimated urine and faecal output (estimated via a survey document) and the authors reported no differences in bwt loss (~4.0%) nor any apparent effect on performance (speed/placing).

There were significant effects on a number of parameters including plasma Na and Cl concentrations (higher at 80km with the higher dose but no difference in early recovery) as well as HCO3 and pH (lower throughout the ride and in early recovery with the higher dose). Although the authors reported no significant change in K there was a trend for the values to be higher with the high dose. Cortisols were higher with the high dose whilst PCV and Hb were lower.

Practical interest

This study supports the belief that increasing salt intake increases the water intake but suggests that at least for 80km rides under moderate conditions giving higher amounts of electrolytes may not have a beneficial effect on performance compared with giving lower levels of supplementation.

However, with both levels of supplementation there was a trend for an increase in chloride concentration and no evidence of a metabolic alkalosis. This may have attenuated the hypochoramia and metabolic alkalosis that may be associated with certain endurance rides. This may be beneficial (especially during longer or more strenuous rides), although the authors caution that efficacy and safety of the higher dose needs to be determined.

Comments on the study

- Some positives:
  - Good field study design.
  - Cross over using the same animals over the same course under similar environmental conditions.
  - Well presented data.
- Some concerns:
  - *Study design issues*: Small numbers of animals; Estimated water, faecal and urine amounts rather than actual; Dietary electrolyte intakes not provided – prior to and during the ride; Lack of any control group without any electrolyte supplementation; Timing of the preride blood samples relative to electrolyte supplementation unclear.

— *Safety issues*: Authors do not provide any real indication as to why they would be concerned with the safety of the higher dose.

— *How to use:* Perhaps most important is the question: Should this study be translated to the endurance rider and if so how - and how would they interpret the paper without being provided with any additional information or are just given a 'sound bite' of the results? – I.e. does this support them giving electrolyte supplementation and if so how much. How can this be translated into practical advice in different rides in different countries?

It is frustrating that although this well designed field trial gives tantalising information it does not really provide any answers, for example:

- Can we or should we interpolate these results to rides over longer distances and more adverse environmental conditions - what about shorter rides?
- Can we be sure that a significant finding in such a trial is of importance in an individual competing horse?
- Can a non significant finding or trend in such a study that may be overlooked or considered of little value actually be of significance to an individual under certain circumstances?
- What can we take from the fact that there was apparently a greater water intake, yet no difference in estimated urine and faecal output and no appreciable difference in bwt loss with the high dose? The authors suggested that this could be related to differences in sweating and respiratory water losses - but is this true? Could it be that estimated intakes and losses were inaccurate – or that the different groups used their gut reservoir differently or are there other possible explanations?
- What about the core diets of these animals – how much could they have influenced the extent of the GIT water and electrolyte reservoir and how available is this really?
- When do bwt losses become significant – what do we really want to achieve with electrolyte supplementation?

What next

There really does need to be a concerted effort by the research community to:

- Agree what information we can provide today with respect to electrolyte supplementation.
- Determine where are the major gaps.
- Carry out appropriate studies that can be compared and combined in-order to fill these gaps.

But I think there has to be some recognition that it is unlikely that we will make substantial progress unless large research rides are held under different environmental conditions. But even then it is unlikely that we will ever understand fully this complex area.

Kellon, E.M., 2006. Use of the herb Gynostemma pentaphyllum and the blue green algae Spiruliina platensis in horses. In Proceedings of the 2nd European Equine Health And Nutrition Congress: 50-59.

Objectives

To describe two 'natural' therapies and the author's experience with them in horses.

Why chosen

There has been a marked increase in the number of specialised dietary supplements on the market, unfortunately many substances are marketed without adequate understanding of their function *in the horse*; too often there is little or no evidence that the metabolic or physiological mechanism, which they contribute to, is a limiting factor or influences the particular problem in the first place; nor whether other processes may be influenced, and no real evidence that they will affect the actual problem or improve performance or behaviour in the field. Very few have any proven science *in the horse* to back their product claims and often there is little appropriate safety data available. There is also a potential risk that some of these products, especially when untested, might alter the way that veterinary medicines act, which may pose an additional health and welfare risk. This paper was chosen as it dealt with some of these issues. In particular I will concentrate on the claims made around the use of Gynostemma in laminitis.

Background

Many people today in developed countries add supplements in order to personalise the diet, add that little bit extra to boost performance, improve health, correct an imbalance etc. Some of these supplementary feedstuffs that we add today have been included in the diets of horses for centuries: Garlic being one such example *'Before 1733 the diet [of racehorses] consisted of barley, beans, wheat-sheaves, butter, white wine and up to 25 cloves of garlic per feed'.* However, the feeding practices of adding ale and dairy products is perhaps less prevalent today than in the 1700s when *'new strong ale and the whites of 20 eggs or more, with no water'* was recommended. In fact we now are aware that there are potential negative consequences to both the feeding of large amounts of garlic and egg whites - confirming that not all 'traditional' practices are safe or beneficial.

The addition of herbs (plants grown for culinary, medicinal, or in some cases even spiritual value), to equine diets does however, appear to be on the increase. They are commonly included for their aroma and perceived palatability as well as their potential medicinal value. Many potential claims for the health benefits of herbs have been made for man and the horse, in particular with respect to respiratory and digestive problems as well as for pain management and improved performance. There has, however, been very little published research on the medicinal effects of herbs in horses, as well as the short and long term benefits OR concerns (especially safety) of herbal preparations.

Overview of part of the paper

Part of this paper describes the herb Gynostemma pentaphyllum which comes from southern China and other parts of Asia and has been known to have been used as a medicinal herb since around the 14th century. The active components are given as being flavones and gypenosides and the author lists the potential effects of this herb that might be of interest to the horse in particular its antioxidant, nitric oxide modulation and bronchodilation activities. Some data to support these effects in man and other species is provided. There is also some discussion of the potential benefit of this herb being an 'adaptogen' which the author describes as meaning that it has the potential to modulate the response to stressors such as disease, infection or exercise. With respect to laminitis – the author suggests that the postulated anti-inflammatory/antioxidant and oxygen free radical scavenging properties of the herb plus in particular its potential to modulate nitric oxide production may be of benefit.

There is some discussion of the potential risks of using this herb especially concurrently with other medicines that have a vasodilatory and/or hypotensive action (Acepromazine) or other drugs/herbs that might have an anticoagulant effect.

Main findings

The author reports on her experience of using the herb in animals with chronic laminitis known to be hyperinsulinaemic, but with normal blood sugars. Based on specific inclusion criteria (including a history of ongoing lameness for at least 6 weeks) around 176 horses are reported as having been monitored –75% of which were on NSAID therapy for pain control prior to the study. Horses were fed from 1-2g of the Gynostemma powder/500lbs twice a day. It is reported that ~62% of animals returned to pasture soundness at a walk within 2 days to 2 weeks of starting the supplementation; ~11.5% were non responders and the remainder showed improvement of 1 – 2 lameness grades.

Practical interest

Veterinarians should always check what herbs or other nutraceuticals animals in their care may be supplemented with, as they may affect the action of any medicines they administer.

A watching brief should be set on this herb – but currently there is insufficient evidence available to support its routine use in laminitics.

Comments on the study

- Some positives:
  - Some reference is made to published scientific work, which supports the broader claims and starts to provide information and support re potential modes of action.
  - Reference to potential toxicity especially with respect to concurrent use of other drugs and herbs.
  - Provides the inclusion criteria used and relatively large numbers of animals are evaluated.
  - Mentions the issue re variation in active ingredients in different sources of the herb and problems of dose determination across species.
- Some concerns:
  - *Insufficient detail*: This is a general review and therefore detailed information is unlikely to be provided and it is not – this does make it difficult to judge the efficacy of this herb and its likely benefit especially as there is no reference to a published paper which provides the details. The chronic laminitis group evaluated seem to include a mixture of animals with and without PPID (pars pituitary intermedia dysfunction). There is no indication of who is providing the observational data and how 'more mental alertness' was judged and exactly what is meant by 'increased spontaneous movement'. What scale was used to judge the 'pinker color to the oral mucus membranes and tongue' especially as this was used to judge when a therapeutic dose had been reached. What happened when this dose was reached – and how were the animals introduced to the herb etc.? Who evaluated the lameness grades and what scale was used – was there any correlation between initial lameness type and outcome?
  - *Study design issues:* No information is provided as to why these animals were so lame. No 'control' group or alternative treatment group were used for comparison – without more information it is difficult to determine if some of these animals would have improved regardless. Additional clarity regarding the inclusion criteria is needed especially as the author mentions using it in racehorses. No mention of any follow up blood samples – do the animals have to remain on the herb and does their improvement continue – were any able to return to work? Was there any

change in Body condition score that might have contributed to an improvement? How long did they need to be on the low NSc diet etc. (again may have contributed).

— *Basic concept*: The mode of action relies heavily in this review on the importance of nitric oxide. However, there does not appear to be any evidence of a deficiency in intrinsic nitric oxide production in laminitis and the jury is out on the importance of this system in laminitis per se – especially chronic laminitis. The rationale for using a nitric oxide modulator in chronic laminitis was not clear. But this does not mean that the herb does not have a role to play – just that it may not work through this mechanism. Evidence is needed to support the comments that the antioxidant action of the herb etc may have a role to play in laminitis.

— *Potential negatives in laminitis specifically*: Although no adverse effects were reported, no balanced discussion was held re some of the other potential effects of this herb that might not be of value in insulin resistance laminitics or those on the verge of being laminitic – for example an extract of this herb has been shown to increase the insulin response to glycaemia in rats (improving glucose tolerance: Norberg *et al.*, 2004). However, this may not be advisable in horses as this may in turn potentiate any negative effects of insulin on the vasculature.

It is important to realise that within the human field a lot of work is currently being undertaken into the potential active ingredients present in this herb for example six dammarane glycosides, plus rutin, kaempferol, quercetin and linalool 3-0-β-D-glucopyranoside were identified in one study reported in 2006 (Yin et al., 2006). There also has been some work done on toxicity with no side effects in rats being seen when fed 8g/kg orally for one month (total gynosaponin) or the water extract at 750mg/kg BW orally for 24weeks. The LD50 of the gynostemma water extract given intragastrically was 4.5g/kg (see Razmovski-Naumovski et al., 2005).

What next

Currently the lack of information on the use of herbs and other nutraceuticals leaves anyone involved with horses with a dilemma – 'how to avoid dismissing possibly advantageous supplements that will add to the quality of life of horse and rider because of a lack of evidence *of efficacy, but without accepting all 'claims' that lack any supporting evidence.'*

This paper provides some tantalising information but it is impossible to judge the real benefit of this herb from this review – *This author should be encouraged to write these cases up as an evidence based medicine case report and subject the information to peer review.* It is essential that this occurs so that we do have information on which we can base our therapeutic decisions.

References

Norberg, A., Hoa, N.K., Liepinsh, E., Van Phan, D., Thuan, N.D., Jornvll, H., Sillard, R. and Ostenson, C.-G., 2004. A novel insulin releasing substance, phanoside from the plant Gynostemma pentaphyllum. Jo of Biological chemistry 279 (40) 41361 – 4167.

Razmovski-Naumovski, Huang, T.H.-W., Tran Van Hoan, Li, G.Q., Duke, C.C. and Roufogalis, B.D., 2005. Chemistry and pharmacology of Gynostemma pentaphyllum. Pytochemistry reviews 4 197 – 219.

Yin, F., Zhang, Y., Yang, Z., Cheng, Q., and Hu, L., 2006. Triterpene Saponins from Gynostemma Cardiospermum. J Nat Prod. 69 1394 – 1398.

Bailey, S.R. and Harris, P., 2006. Effect of dietary fructan carbohydrates on plasma insulin levels in laminitis-prone ponies. Journal of Veterinary Internal Medicine May/June 2006: P799-800.

Objectives

To assess the insulin responses to dietary fructan consumption in non-obese ponies predisposed to laminitis, compared with unaffected controls.

Why chosen

Laminitis occurs all around the world in horses as well as ponies and has major welfare implications. It is obviously important to be able to recognise and treat the condition in its early stages so that the pain and suffering is kept to a minimum. However, ideally we would like to be able to recommend certain interventions/ countermeasures that would help to avoid or prevent the condition from occurring in the first place. Obviously as pasture associated laminitis occurs at pasture then the easiest way to avoid the condition is to prevent access to pasture and to feed forage alternatives that are known to be low in rapidly fermentable material. For the majority of horses total restriction is not always a viable or desired option for financial, welfare and health reasons. It also may not be necessary for those animals that are not predisposed to (i.e. have an increased risk of developing) laminitis. It is therefore increasingly recognised that identifying individual animals at risk, before they suffer serious episodes of laminitis, will potentially be of great benefit; bearing in mind the difficulties in treating and managing recurrent and chronic laminitis. This abstract illustrates one potential method to do this.

Background

Laminitis is perhaps most commonly associated with certain feeding and managemental factors that will increase the likelihood of a potential attack whatever the type or breed of horse. Turning certain ponies out onto lush pastures in the spring and autumn is a common triggering factor for the development of laminitis. Currently it is thought that the high levels of water-soluble carbohydrates (which include the simple sugars as well as the more complex storage carbohydrate: *Fructans*) and/or starch may be involved in this process. It is thought that as for other mammals the horse does not have the necessary enzymes to digest fructans directly within the small intestine. Fructans therefore pass relatively unchanged into the hindgut where they are readily fermented, in a similar manner to starch that escapes digestion in the small intestine following the ingestion of too large a cereal based meal. Rapid fermentation will lead to the swift production of lactic acid in particular, and the lowering of the pH upsetting the bacterial / microbial balance. When this occurs and the pH drops below a certain critical point, bacteria that are not able to survive under such conditions may die and release *endotoxins* and other such compounds into the hindgut. In addition there may be overgrowth of certain bacteria that can survive in such conditions which in turn release/produce other compounds into the large intestine. Changes in the mucosa due to the drop in pH and other factors may make it more permeable to such substances. A number of compounds, may be absorbed into the blood in greater amounts than normal and have further effects, in particular within the feet (not necessarily directly), triggering the development of laminitis. These include the *vasoactive amines* that may potentiate a reduction in the blood flow to the feet and activators of various *Matrix metalloproteinases*, which may potentiate a more rapid and unwanted separation of the basement membrane between the lamellae. These effects may be potentiated by the animal being *insulin resistant*.

Insulin resistance is increasingly being linked with an increased risk of laminitis and in particular is thought to be a major component of the prelaminitic metabolic syndrome. Currently the most accurate methods of assessing insulin sensitivity are the euglycaemic hyperinsulinaemic clamp technique and the insulin-modified frequently sampled intravenous glucose tolerance test with minimal model analysis of the data, neither of which are suitable for use in practice. Single basal insulin and glucose samples are not indicative of insulin resistance, as many insulin-resistant ponies have basal insulin concentrations well within the normal range. However, proxy values derived from basal insulin and glucose may have some predictive value if considered together with other factors (Treiber *et al.*, 2005). More simple dynamic methods of determining differences in insulin response may be of additional value.

Overview of the abstract

Eleven adult non-obese mixed native breed ponies (4 mares and 7 geldings) were used. Six of the ponies had one or more episodes of acute laminitis in the previous two years, but none had shown clinical signs in the 3 months leading up to the study (laminitis-prone group); the remaining five had not shown any signs of laminitis in at least the previous 3 years (normals). The ponies were fed hay *ad-libitum* for two weeks, before inulin (a commercially available form of fructan carbohydrate; 3g/kg) and dried grass (Readigrass®; replacing one third of forage ration by weight) were included, split into three meals. The amount of inulin in the diet was safely below the amount expected to cause laminitis, based on published data from other groups using raftilose, and caused a small but significant decrease in faecal pH (from 6.8 ± 0.1 to 6.3 ± 0.1 pH units); no clinical signs were observed.

Patricia A. Harris

Blood samples were taken (3 hrs post-prandial) for insulin, glucose and triglyceride analysis, before and 1, 2 and 5 days after the inclusion of inulin.

Main findings

No significant changes in plasma glucose or triglyceride levels were observed in either group following the addition of fructan carbohydrates to the diet. In the normal group of ponies, insulin concentrations increased by 2.0±0.3 fold between diets (median on hay diet 14.4 µIU/ml vs. 37.9 µIU/ml on hay plus inulin, 48hrs (no significant difference). Normal range 5.5-36.0 µIU/ml. However, in the ponies predisposed to laminitis, although their insulin levels were not significantly different from the normal ponies while on the basal hay diet (44.7: 3 within normal range), insulin concentrations increased dramatically and significantly (5.5±2.2 fold increase) following the addition of fructan carbohydrates with a peak of 137.0 (range 52.1-576.0) µIU/ml at 48 hrs.

Practical interest

The ponies predisposed to laminitis are likely to have compensated insulin resistance, which may not be apparent when plasma insulin concentrations are measured during the feeding of diets containing low amounts of rapidly fermentable carbohydrate. However, in these individuals, dietary fructans are capable of unmasking an exaggerated insulin response. This effect may be useful in the identification of animals suspected of being insulin-resistant and therefore predisposed to laminitis. It also has implications for the feeding and management of such animals.

Comments on the study

- Some positives:
  — Well characterised ponies.
  — Good study design.
  — Although it does not necessarily indicate insulin resistance, this test may be of some practical value in demonstrating an abnormality in insulin secretion. It may also explain why ponies

are at risk when fructan carbohydrates are increased in spring/summer grass.

- Some potential concerns:
  - *Insufficient detail*: This is an abstract and therefore full details are not available.
  - *Study design issues*: No other method of evaluating their level of insulin responsiveness was carried out. Small numbers, single dose. Needs to be repeated in a different group of animals.
  - *Potential negatives:* are there any risks of undertaking this test in animals for example on the cusp of an episode – could it tip them over? What does it really indicate?
  - *Practicality*: How practical is this test?

What next

Further evaluation of this technique in other groups of well-characterised animals is required in order to confirm the value, determine optimal dose and most importantly confirm whether this can be a useful test in determining an increased risk in animals not previously known to suffer from this condition.

Reference

Treiber, K.H., Kronfeld, D.S., Hess, T.M., Boston, R.C. and Harris, P.A., 2005. Use of proxies and reference quintiles obtained from minimal model analysis for determination of insulin sensitivity and pancreatic beta-cell responsiveness in horses. Am J Vet Res; 66:2114-2121.

Evaluating recent equine research (2005-06)

Brian D. Nielsen

Dept. Animal Science, Michigan State University, East Lansing, MI 48824-1225, USA

For the average person, the most difficult part of nutritionally managing their horse can be figuring out what feeds or supplements are advantageous to feed their horse and which ones might be a waste of money. The majority of individuals rely on advertising (which obviously can be misleading), articles in popular press magazines (which often do a good job of evaluating research but sometimes don't), or their local veterinarian (some of which have a nutrition background, but many do not, and they may find it difficult to stay current on research findings due to their busy schedules). Ideally, horse owners would read peer-reviewed research articles themselves to determine whether certain products or feeding strategies actually have merit. However, doing that is both impractical and may result in greater confusion as, to properly review research studies, one has to be familiar with scientific terms and also needs to be able to recognize limitations in research that might limit the conclusions that one can draw from even peer-reviewed and published research. The four papers to be discussed all have appeal to various segments of the equine industry. However, as will be pointed out, there are limitations to the findings of these papers despite them being published in peer-reviewed journals. These limitations do not necessarily negate the findings, but they do prevent one from making too strong of conclusions about the item being tested.

Inoue, Y., Asai, Y., Tomita, M., Kuribara, K., Kobayashi, M., Kaneko, M. and Toba, Y., 2006. The effect of milk basic protein supplementation on bone metabolism during training of young thoroughbred racehorses. Equine Vet. J. Suppl. 36:654-658.

Objectives

To evaluate the effect of milk basic protein supplementation on bone metabolism in young Thoroughbred racehorses in training.

Why chosen

The results of this study, when initially reviewed, appeared to be almost 'too good to be true.' If future research continues to support the findings of this study, the implications would be dramatic.

Background

Finding ways to improve skeletal strength and, thus, likely decrease the incidence of injuries to horses is a focus of many researchers and a desire of those who work with athletic horses. While one primary avenue to doing this is through proper exercise, finding a nutritional approach is appealing.

Overview of the study

Twenty 2-year-old Thoroughbreds were used in this 90-d training study. The treatment group was fed a basal diet with 1 g of milk basic protein and the control group was fed a basal diet only. Blood samples were taken at the start of the study, on d 45, and on d 90 to determine serum calcium and biochemical markers of bone metabolism. Radiographs were taken at the start and end of the study to determine radiographic bone aluminum equivalence (RBAE) as a measure of bone mineral content.

Main findings

The treated group had higher concentrations of osteocalcin (a marker of bone turnover – often used as an indicator of bone formation) at d 45 compared to the control group. The treated group also showed a greater percent increase in the total RBAE by the end of the study compared to the control group. There were no differences in serum calcium or carboxy-terminal telopeptide of type I collagen between groups.

Practical interest

If the findings are correct, this represents a very simple management tool to increase bone formation and bone mineral content.

Comments on the study

It is surprising that such a small amount of a substance (1 g/d) could have such a dramatic impact on bone. Typically, rather small differences in nutrition do not make a *measurable* difference on bone quantity when nutritional requirements are being met. However, the authors indicated that it takes 20 L of milk to obtain 1 g of milk basic protein so, theoretically, 1 g could represent the impact that could be made by the consumption of a much larger quantity of milk – at least in terms of milk basic protein. While there has been some research into the role that milk basic protein plays on bone metabolism, this area of research seems relatively new and it is expected that more research in this area will continue.

While there is little to criticize about the research, the results may not be as dramatic as presented. Due to some variation in RBAE at the beginning of the study, the authors chose to evaluate RBAE by examining the percent change from the starting value. From a practical standpoint, that is reasonable. In horse studies, we typically are forced to work with small numbers of animals and a little variation in starting values can influence whether differences are found at the end of the study. However, from a strict statistical standpoint, it could be argued that making such transformations of data might not be appropriate. Given that the starting values were not different ($P = 0.18$), one could argue that it was not appropriate to transform the data to account for 'differences' in starting values. Given that the ending values were also not different ($P = 0.45$), a purist in statistics would likely say that the treatment had no effect on bone mineral content. That being said, and recognizing the limitations that exist in conducting horse research, I would argue that the transformation the authors made was reasonable in helping to detect a potential effect. However, given that a true difference was not detected by treatment, this study should only be considered evidence that administration of milk basic protein may help enhance the quantity of mineral present but it does not conclusively prove that it does. The authors, too, suggest that further studies would be required to substantiate the findings.

When possible, one technique to help eliminate the variation in data between treatment groups at the beginning of the study is to analyze initial samples immediately and then use those results in helping to stratify the animals, pair-match them, and then randomly assign them. However, that is not always possible and the authors were reasonable in believing that having ten animals per treatment should have taken care of the variation.

One other item should be mentioned. The paper suggests that milk basic protein enhances bone formation (osteoblastic activity) and either does not change or reduces bone resorption (osteoclastic activity). If that occurs, as the findings of this finding support, a subsequent increase in bone mineral content likely could occur but such an increase might be accompanied by a decrease in bone quality and a potential increase in injuries. The process by which bone repairs itself involves both osteoblastic and osteoclastic activity to replace damaged or old bone. If osteoclastic activity is reduced, then it is likely that the quality of bone may be reduced as damaged material is left in place. Thus, simply increasing the amount of mineral present may not reflect a bone better prepared to handle the rigors of racing and could potentially result in a greater injury rate. While there is no guarantee that such is the case, future studies would need to examine this to be certain that feeding milk basic protein is not setting a horse up for injury.

Forsyth, R.K., Brigden, C.V. and Northrop, A.J., 2006. Double blind investigation of the effects of oral supplementation of combined glucosamine hydrochloride (GHCL) and chondroitin sulphate (CS) on stride characteristics of veteran horses. *Equine Vet. J. Suppl.* 36:622-625.

Objectives

To quantify the effects of an oral joint supplement (combination glucosamine hydrochloride (GHCL), chondroitin sulphate (CS) and N-acetyl-D-glucosamine) *in vivo* on stride parameters of veteran horses.

Why chosen

This paper is an example of investigators trying to perform research in an ideal way (blinded and controlled) but that failed to striate the horses in an appropriate manner before randomly assigning them to a treatment group – resulting in results that could be called into question.

Background

Chondroprotective agents such as glucosamine and chondroitin sulphate are recommended by many veterinarians and purchased by many owners for the purpose of preventing or eliminating the symptoms of osteoarthritis. This is despite the lack of strong evidence to show that they have any measurable effect.

Overview of the study

Twenty veteran horses were randomly assigned to a treatment (n = 15) or placebo group (n = 5). Pre-treatment gait characteristics were recorded at trot using digital video footage. The range of joint motion, stride length, and swing and stance duration were assessed using 2-dimensional motion analysis. Treatment (or placebo) was administered daily for 12 weeks at the manufacturer's recommended dosage. Gait was reassessed every 4 weeks using the pre-treatment protocol. Researchers and handlers were blinded to treatment throughout the study.

Brian D. Nielsen

Main findings

By week 8, treated horses had increased range of motion in the elbow, stifle and hind fetlock and the stride length had also increased.

Practical interest

There has been a need for a well-controlled and blinded study evaluating the efficacy of chondroprotective agents such as these. The most commonly cited study that showed an increase in stride length had no controls. Despite numerous *in vitro* studies giving reasons why they might work, only a few studies have shown any results suggesting glucosamine and chondroitin sulphate actually do work. A federally funded project (versus one funded totally or in part by a company producing or selling these substances) was published last year in the New England Journal of Medicine. With 1,583 patients divided into four treatment groups (placebo, glucosamine, chondroitin sulphate, glucosamine & chondroitin sulphate combined), it was determined that neither glucosamine nor chondroitin sulphate by themselves showed any effect. With the combination of glucosamine and chondroitin sulphate, 66.6% of the patients reported improvements. However, this was considered to be a trend at $P = 0.09$ as the patients in the placebo group reported improvements at the rate of 60.1%. This study demonstrates the potential magnitude of the placebo effect and the importance of conducting controlled studies. It also demonstrates the difficulty in demonstrating efficacy of these substances suggesting that if they do work, their impact is relatively minor given the difficulty in detecting differences using a very large sample size.

Comments on the study

While the authors are to be commended in trying to set up the project correctly, the first major concern was allotting only 5 of the 20 horses to the control group. The reason for this is an alleged variability in the uptake of glucosamine and chondroitin sulfate. Even if such variability exists, it is reasonable that this is part of the factor in determining whether it is efficacious or not. If it is relatively unabsorbed by a number of horses, that lessens the practical use of it and it should be accounted for in a balanced study if the effect is great enough. If the effect is not great enough, it begs the question as to whether recommending the usage of these products at a substantial cost per day is warranted. Additionally, regarding variability, the authors mentioned that the horses ranged in age from 15 to 35 years (a 20-year difference) and were of varying heights and breeds. This is certainly reasonable and expected when trying to find sufficient numbers of horses of that age that can be used in the study, but such variation in the animals would require that that they be striated according to age, height, and breed as all of these certainly would have an effect on joint motion. Likewise, similar to what was mentioned in the previously mentioned study, it would have been ideal to use the pre-trial joint characteristics as another factor in the striation process. This is especially true given the control group only contained five animals. To further complicate matters, though it was not mentioned in the paper, I believe horses were housed at three different locations so were under different management. Given all the variation in age, height, breed, joint characteristics, and location, and given the small size of the control group compared to the treated group, one thing that could have strengthened the paper is information as to the description of the horses in each group. If each group had similar horses and if the groups were distributed equally between the three locations, it strengthens the results of the study. If not, the results can certainly be called into question. Regardless, studies like this are definitely needed to ascertain whether these products are efficacious and the authors are to be commended for trying to determine such.

Brinsko, S.P. Varner, D.D. Love, C.C. Blanchard, T.L. Day, B.C. and Wilson, M.E., 2005. Effect of feeding a DHA-enriched nutriceutical on the quality of fresh, cooled and frozen stallion semen. Therio. 63:1519-1527.

Objectives

To determine if feeding a nutriceutical rich in docosahexaenoic acid (DHA) would improve semen quality.

Why chosen

This study appears to be strong with only one item for which further information would have been useful. However, this study is being discussed to caution against over-extrapolating the results of a single study.

Background

The incorporation of omega-3 fatty acids into diets of various mammalian species is gaining in popularity for various reasons including cardiovascular health, performance enhancement, and, as indicated by this study, enhanced reproductive performance. A number of studies are being conducted in the equine though only a few have made it through the peer-review process and have been published.

Overview of the study

Eight stallions were used in a 2 x 2 crossover study in which stallions were randomly assigned to one of two treatment groups (n = 4 per group) with groups balanced based upon semen analysis from baseline measurements. Stallions were fed their normal diet (control) or their normal diet top-dressed with 250 g of a DHA-enriched nutriceutical. Feeding trials lasted for 14 weeks, after which a 14-week washout period was allowed and the treatment groups were reversed for another 14-week feeding trial to allow for determination of supplementation on semen quality.

Main findings

Feeding DHA resulted in a three-fold increase in semen DHA concentrations and a 50% increase in the ratio of DHA to DPA (an omega-6 fatty acid) in semen. Sperm motion characteristics in fresh semen were unaffected by treatment but after 24 hours of cooled storage, sperm from stallions fed the nutriceutical exhibited higher velocity and straighter projectory. After 48 hours of cooled storage, increases in the percentages of sperm exhibiting total motility (P = 0.07), progressive motility (P = 0.06) and rapid motility (P = 0.04) were observed when horses were supplemented. Feeding the nutriceutical also resulted in improvements in mean progressive motility of sperm after 24 hours for a subset of four stallions whose progressive motility was < 40% after 24 hours of cooled storage when fed the controlled diet.

Practical interest

This study demonstrates that the omega-3 fatty acid DHA can positively impact stallion semen after cooling. For stallions whose semen ships poorly, supplementing DHA in the diet may be a simple way to improve semen quality.

Comments on the study

Only one concern, and likely it is a minor one, exists with this study. The experimental design called for eight horses to be used in a 2 x 2 crossover study which would result in four stallions being initially on the control diet and four stallions on the DHA-supplemented diet. Though the authors indicated that the stallions were randomly assigned to treatment groups after balancing for stallion age and semen quality, no assurance was given indicating that examining semen after cooling for 24 and 48 hours was part of the evaluation of semen quality. Thus, there is the potential that the four stallions that would be considered 'poor shippers' were not equally divided between the two groups. If they were not equally divided, the improvements seen in semen characteristics after cooling may have more to do with a seasonal effect rather than the supplementation given that the study began in November and ended in October of the following year. As changes in daylight length have an effect on fertility, reassurance that those four stallions were equally divided between groups, if that were the case, would have been useful to eliminate concerns. If they were not equally divided, this issue should have been addressed. Again, this is not a major concern but it is one area in which some more information would have been useful for readers.

A bigger concern deals not with this paper, but the potential of a reader over-extrapolating the results of this study to other scenarios. Given that this study demonstrates some likely improvements in fertility associated with supplementation of DHA, the potential exists for one to believe supplementing omega-3 fatty acids would also improve fertility in mares. However, a study presented in 2006 by

Poland *et al.* reported that mares supplemented with both DHA and EPA during late gestation and early lactation had the first post-partum ovulation delayed by about 10 days. Additionally, treated mares held a large follicle (≥ 35 mm) during the first postpartum estrous period for over 6 days longer. Supplementing with just DHA did not result in such delays. However, there are enough sources of omega-3 fatty acids that include both DHA and EPA that one needs to be aware of this concern before supplementing mares in the belief that omega-3 fatty acid supplementation would likely have a positive effect on fertility similar to what was reported with stallions. If mares are supplemented with these omega-3 fatty acids, the strong potential for reduced fertility exists.

References

Poland, T.A., J.M. Kouba, C.M. Hill, C. Armendariz, J.E. Minton, and S.K. Webel. 2006. Effects of fatty acid supplementation on plasma fatty acid concentrations and characteristics of the first postpartum estrous in mares. J. Anim. Sci. 84(Suppl. 1):393.

Turner, K.K. Nielsen, B.D. O'Connor, C.I. and Burton, J.L., 2006. Bee pollen product supplementation to horses in training seems to improve feed intake: a pilot study. J. Anim. Phys. Anim. Nutr. 90:414-420.

Objectives

To determine the efficacy of supplementation of Dynami Trio 50/50, a bee pollen-based product, to improve physical fitness, blood leukocyte profiles, and nutritional variables in exercised horses.

Why chosen

Many supplements have been sold to the equine community for years without any testing to determine efficacy in horses. While testimonials often exist, scientists generally dismiss anecdotal evidence that does not have scientific data to back it. This study examined one such product and actually found some benefit from supplementing it in the diet. However, there are limitations to the study that prevented the potential detection of other benefits (emphasizing the importance of proper experimental design) and the study also was sufficiently small to warrant caution in over-emphasizing the benefit that was detected.

Background

Many individuals in the horse industry feed 'bee pollen' to their horses in the belief that it improves exercise performance and health. While there have been some studies examining feeding 'bee pollen' to other species, namely rats and humans, there apparently had not been any published studies in horses investigating the efficacy of this practice.

Overview of the study

Ten unfit horses were placed into a 42-day training study to evaluate whether feeding a bee-pollen based product could improve athletic performance, immunological parameters, and nutrient digestibility. Horses were subjected to a standardized exercise test (SET) on d 0 to determine pre-training fitness. From that, horses were pair-matched by gender and pre-training fitness, and were randomly assigned to either a treated group which received 118 g of Dynamic Trio™ 50/50 consisting of 55% 'bee pollen' or a control group which received 78 g of a placebo that was similar in volume and appearance to the supplement given the treated horses. All horses were individually fed their treatments once per day mixed with 0.2 kg of sweet feed to ensure consumption but were allowed ad libitum access to roughage. Supplementation continued for the duration of the study. Horses were conditioned through riding, treadmill exercise, and exercise in a free-flow walker with increasing speeds and duration occurring throughout the study. At the end of the study, all horses completed another SET. Blood samples were taken before, during, and after each SET. Other physiological parameters such as heart rate and blood lactate concentrations were also taken. From d 18 to 21, the six geldings underwent a total collection period to examine nutrient digestibility. During that period, horses were given their respective treatment and grain, and continued to have ad libitum access to hay, with the amount consumed being monitored and recorded.

Main findings

While physical fitness increased greatly from d 0 to 42, no treatment differences were noted. There were some changes in immunological parameters associated with exercise, but few treatment differences were observed and it was difficult to determine if the differences that were observed were of any significance given their relatively small degree of change. Surprisingly, there were major differences in nutrient retention and digestibility despite only testing these items in three horses per treatment (only the geldings). Treated horses had less phosphorus excretion and tended to retain more nitrogen. They also tended to digest more NDF and ADF while having lower NDF digestibility and tending to have lower ADF digestibility. Those results (digesting more while having a lower digestibility) did not make any sense until the total amount of feed consumed was examined. It was revealed that the treated horses ate more hay (9.4 $\pm$ 0.3 kg/d) than did the controls (6.3 $\pm$ 0.5 kg/d; P < 0.0001).

Practical interest

On every one of the three days of the total collection portion of the study, all three treated horses ate more hay (the only portion of the diet for which they had free access) than did any of the control horses. Given that hard-working horses are often known to lose their appetite – often resulting in weight loss and reduced performance – this higher rate of feed consumption likely could have an effect on the performance of horses.

Comments on the study

There were several limitations to this study. The first being that the horses had gained sufficient fitness that the test used on d 0, and of which many of the horses could not complete, was insufficient to test maximal effort on d 42. Hence, though we could not detect any treatment difference in the fitness parameters, we could not conclude that the treatment did not have any effect – all we could conclude was that we could not detect any effects using our testing procedure. Likewise, as seen with a number of studies examining immunological parameters, even when differences are noted, it is difficult to conclusively say such differences are positive or negative. In contrast, the finding that there was a difference in feed intake was dramatic and certainly suggest that the supplementation of this product stimulated appetite. One item that would have been nice to know is whether there were differences in body weight loss or gain associated with treatment. Unfortunately, since such differences in feed intake were not expected, body weight measurements were only taken during the total collection period and not before and after the study. If that difference in feed intake persisted through the course of the study, it would have surely resulted in weight differences. Unfortunately, simply testing the product on feed intake in a group of horses not in training may not result in similar findings. One hypothesis as to why feed intake was stimulated is that the product is reported to be high in B-vitamins and a failure to produce enough thiamin has been suggested to be one of the reasons for appetite loss in stressed horses. If that is the case, one would not expect similar differences in feed intake in non-stressed horses. Alternatively, after the paper was published, the authors became aware of some phagostimulant properties for bees that are found in natural pollen but not in synthetic pollen. Potentially it is these properties that may be having a similar appetite stimulant effect in horses, but this is only speculative at best.

In summary, this is a product that I have been skeptical of for years. However, after doing the research, I now believe that it can have an effect in hard working horses. Just how big an effect it is remains to be determined but it does demonstrate a case in which a company was willing to support research testing of their product, knowing that there is a chance no effects will be found, but the research ends up supporting some of their claims. This study also shows that even when an effect is found, there is still plenty of room for additional research to better elucidate the mechanism behind an effect. More companies are challenged to support research of their products to ensure consumers have some evidence that a product is efficacious.

Noteworthy changes to the horse nutrition requirements by the National Research Council of USA

Brian D. Nielsen

Dept. Animal Science, Michigan State University, East Lansing, MI 48824-1225, USA

It has been eighteen years since the National Academies Press published the Fifth Revised Version of the National Research Council's Nutrient Requirements of Horses – know to many as the 1989 Horse NRC. Since that time, a substantial amount of research has been done to better determine nutritional requirements of horses and ways to feed them to improve performance. Though many owners, trainers, veterinarians, and individuals in the feed industry have been anxiously awaiting the publication of the Sixth Revised Version, given that the 2007 publication is over three times longer than the 1989 version, it is doubtful that many people will take the time to read the entire publication. While a number of requirements or feeding suggestions did not change dramatically, there are some differences worth highlighting – especially some of those that will impact individuals feeding equine athletes and growing horses.

Probably the item that will have the biggest impact on the way people think about feeding horses is the change from discussing nutrient requirements on a concentration basis (i.e. a horse needing 10% crude protein) to expressing the requirement on a body weight basis (i.e. 0.266 g crude protein per kg of body weight or 0.0043 ounces per pound of body weight). While this may seem intimidating to a lot of people, it is actually quite simple to determine a horse's requirements (simply multiply the requirement by the weight of your horse) and this emphasis on total amount consumed per day versus the concentration in the feed was done for several reasons. First, as many horses' diets, especially those formulated for equine athletes, now contain added fat, the total amount consumed by the horse likely will go down as fat provides more calories in a smaller amount. As a result, a horse on a high fat diet will eat less feed and, if one is feeding nutrients on a concentration basis, the total amount of the nutrient would also decrease. By feeding on a body weight basis, the amount of a nutrient will remain consistent regardless of the total quantity of feed fed.

Another issue is that many individuals only consider the concentrate or grain portion of the diet when evaluating their feeding program. When asked what one feeds their horse, most individuals will respond with an answer like, 'I feed a 12% sweet feed.' The error in this is that the grain only constitutes part of the ration, sometimes a very small part, and the remainder of the diet (usually hay) is often ignored. Additionally, using protein as an example, even if the person is feeding a 12% crude protein grain mix, the percent protein in the total ration may far exceed 12% (e.g. if they are feeding a high quality legume hay that is 17% or greater crude protein) or it might be far less than 12% if feeding a low quality hay. Now when analyzing what one is feeding, it will be much more logical to consider the contribution of nutrients from all of the feedstuffs and not just the grain.

Another major improvement to the new Horse NRC is a web-based computer program that can be used to determine the requirements of horses of varying sizes and uses. The advantage to having it web-based is that no special software, other than a browser, needs to be installed on one's computer, which greatly increases the accessibility of it. While one limitation of many ration evaluation programs has been that they only work with a PC format, this program has been tested to work with Macintosh computers also. The program will allow the user to input many different variables to calculate the requirements such as weight and age, and other criteria such as whether it is a gestating or lactating mare, or if it is a horse in training. Likewise, though a data base with typical nutrient contents of various feedstuffs is included in the program, users can input their own values if they have the analysis of their feeds – thus allowing them to fairly accurately determine if they are meeting the requirements of their animals. On should be able to get access to this computer program by going to the website of the National Academies Press (www.nap.org) and typing 'horse' into the search engine.

The variables of age, weight, use, growth rate, and many others influence the requirements of horses and attempts have been made to quantify the changes these variables have on the requirements. One major change in the 6th edition of the Horse NRC is the recognition that horses vary in their energy requirements based upon temperament and basal activity levels. As a result, the energy requirements for maintenance of a mature horse range from a minimum value that might be appropriate for a docile horse that has little voluntary activity. An elevated value was derived for horses with nervous temperaments that exhibit high degrees of voluntary activity. An average value is suggested for horses that perform moderate amounts of voluntary activity. While it will be up to the horse owner to decide which category their horse might fall into, this range is provided to account for those horses that would be considered an easy- or hard-keeper.

While that change accounts for differences in maintenance requirements (in other words, mature horses that are not being used for breeding and are not in training), another change was to alter the categories used to define the intensity of exercise for working horses. While there previously were three categories of workload (light, moderate, and intense), the new NRC provides four categories (light, moderate, heavy, and very heavy) with each category recognizing an increase in digestible energy requirements (20, 40, 60, and 90% above maintenance, respectively). Obviously, it will be up to the horse owner or trainer to decide which category best describes their horse based upon distance routinely worked, speed, and other conditions, but, fortunately, the evaluation of energy requirements can fairly accurately be done visually. If the horse has adequate fat cover over the body and is maintaining a constant body condition score without losing or gaining much of that fat cover, the energy requirement is being met.

In contrast, it is not nearly as easy to tell if the requirements of other nutrients are being met. Protein, for example, is needed by the horse in order to provide amino acids. In essence, the requirement that the horse has is actually not for protein, but for the individual amino acids, which serve as 'building blocks' that are assimilated into proteins. Thus, in the ideal world, the requirements for the individual amino acids would be known as they are for other animal species. Unfortunately, the only individual amino acid requirement that is known is the one for lysine – the first limiting amino acid (the one that typically is first to limit the formation of proteins when inadequate amounts of it are not present in the diet). Instead, protein requirements are still expressed in terms of crude protein. Crude protein does not even actually measure protein, but instead measures the amount of nitrogen present, which is then multiplied by 6.25. While the new NRC discusses the usage of digestible protein and the advantages of it for better determining protein requirements, sufficient data were not available to allow the committee to satisfactorily suggest digestible protein requirements. The advantage to moving to a digestible protein system is that it better reflects differences in digestibility and amino acid profile of various protein sources.

Like with energy, maintenance requirements for crude protein also range from minimum to elevated and four categories for intensity of work have also been provided. Besides accounting for protein needs for incorporation into tissues, these categories include an allotment to account for protein losses that occur through sweat.

The increase in the use of fat in the diets of horses necessitated a much larger discussion on the subject than what was provided in the 1989 NRC. Some of the conclusions that have been drawn from the research that has been completed since that publication include the need for an adaptation period of roughly three weeks when introducing fat into the diet before the horse experiences some of the advantages of fat in the diet, though two to three months may be needed before complete adaptation to a fat-supplemented diet occurs. Additionally, though there is a fair body of research that indicates supplementing fat to the diet of performance horses improves performance, there is a relatively equal body of research that has failed to detect improvements in athletic performance. Regardless, there are a number of other reasons, besides improving athletic performance, for which the inclusion of fat provides benefits including providing a means to increase the caloric density of a diet without increasing the amount of starch fed, thus decreasing the likelihood of developing problems associated with a high starch diet.

Like with the other nutrients, many of the requirements for minerals were also given as the amount required per day versus as a percentage of the diet. For a few minerals, in particular, some of the trace minerals, insufficient research has been done to alter the requirements given in the 1989 NRC and thus were left on a concentration basis. There has been a substantial amount of research that has focused on skeletal development in young horses entering training and the effect that growth and exercise can have on mineral requirements. While exercise, or lack thereof, can greatly impact the skeletal system, sufficient mineral needs to be present to allow modification of the skeletal system to occur if sufficiently stimulated through exercise. Thus, for the young, exercising horse, the requirement for calcium was increased slightly though this increase was actually accomplished by increasing the requirement for growth. Phosphorus requirements for growth were also similarly increased which should allow sufficient phosphorus for young, growing horses entering training though the requirements for the mature, exercising horse were unchanged from the 1989 NRC. By contrast, the 1989 NRC magnesium requirements were less for growth than for moderate and heavy exercise. While there was no justification for increasing the magnesium requirement for mature horses, the recommendation for young horses under 24 months of age in any intensity of training was established as the same as that for horses in heavy exercise to accommodate potential increased needs associated with the commencement of training. Another interesting change to minerals deals with the electrolytes sodium, potassium, and chloride. Previous estimates from the 1989 NRC for exercising horses were based on level of training (light, moderate, or intense). However, the major limitation to this technique is that vast majority of electrolyte loss associated with exercise is through sweat and using the level of training often may not be closely correlated to sweat loss. By comparison, the new NRC uses sweat loss as a guideline in predicting electrolyte requirements. Directly measuring sweat loss is difficult but the substantial amount of weight loss during an exercise bought is due to sweat loss. Obviously, there can also be weight loss associated with defecation and urination but that tends to be rather small compared to the weight loss associated with sweating during an exercise bout. Therefore, to somewhat accurately predict requirements for these electrolytes, weighing horses before and after exercise would provide the best estimate. It is recognized that many individuals do not have the capability to make such a measurement so an estimate of sweating rate is also provided for each exercise intensity.

The section on trace minerals, like the rest of the NRC, has been expanded and some recommendations were changed. There has been little work specifically looking at cobalt requirements in the horse.

However, given that horses appear tolerant of lower concentrations than cattle or sheep, the requirement was lowered to 0.05 mg of cobalt/ kg of dietary dry matter (also known as 0.05 ppm or parts per million) which should be met through the normal diet of a horse and additional supplementation would not typically be required. Many people have speculated as to what will happen to the requirement for copper. The 1989 NRC requirement was set at 10 mg copper per kg of diet. There was insufficient evidence to suggest that this should be changed for mature, exercising horses. In contrast, the requirements for growth, pregnancy, and lactation have been changed to 0.25 mg per kg of body weight. This assures a minimum of 10 mg copper per kg of diet when a horse is consuming 2.5% of its bodyweight per day but would result in a greater dietary concentration when the horse is eating less than that. For instance, a young, growing horse eating only 2% of its bodyweight per day would be consuming a diet containing 12.5 mg copper per kg of bodyweight. This is not a substantial increase but does ensure that horses eating a smaller amount of feed (for instance, horses on a high fat diet), still receive an adequate amount of copper. While it has been proposed that increasing the copper requirements substantially would prevent many developmental orthopedic diseases, the evidence supporting this is limited. Additionally, while most commercial feeds contain substantially higher concentrations of copper (often 40 to 50 ppm), the incidence of developmental orthopedic diseases (in particular, osteochondrosis) still remains high in segments of the horse industry. This suggests that increasing the requirement substantially is not warranted.

An additional section was added that discussed 'other minerals of interest' - covering chromium, fluorine, and silicon. It was noted that no specific requirement of these minerals by the horse is currently known but there are reasons to believe a true requirement for them may exist. The research on chromium has not conclusively determined a need for supplementation. Additional supplementation is not suggested as any potential requirement would be met by a traditional diet and supplementation may result in problems. Several studies have suggested that silicon in a form that can be absorbed may have some beneficial effects in preventing injuries though it would be difficult to ever prove an absolute requirement due to the relative abundance of silicon found in the environment – even when such silicon might not be easily absorbed by the horse. Boron, nickel, and vanadium are considered an essential part of a mammalian diet but the amount required is very small and supplementing them is not recommended.

Like with many other nutrients, the 1989 NRC expressed the requirements of many vitamins on a dietary concentration basis. They were transformed to a body weight basis in the new NRC. Other than making that transformation, the requirement for vitamin A and vitamin D were unchanged, except for the vitamin D requirement for growing horses, which decreases from a high of 22.2 IU/kg of bodyweight for birth through 6 months and decreases to 13.7 IU/kg of bodyweight for months 19 through 24. While vitamins are an important part of a diet, there is insufficient evidence to suggest a requirement in the horse for vitamin K, niacin, biotin, folate, vitamin B_{12}, vitamin B_6, pantothenic acid, and vitamin C.

Despite the dramatic increase in length of the new NRC, most of the changes to nutrient requirements are not huge. As more research adds to the body of knowledge in horse nutrition, the ability to refine the requirements increases. However, the limitations on conducting horse research (such as small numbers of animals used because of the cost, hesitation to sacrifice animals to look at nutrient deposition in the body, difficulty in detecting small improvements in performance because of the many factors that affect it) make it difficult to be as precise in determining the nutritional requirements of horses compared to other species. Regardless, the emphasis on calculating requirements based on a total amount consumed per day (on a body weight basis) certainly will impact the way many people view feeding horses and will hopefully emphasize the importance of considering the roughage portion of the equine diet and not just the grain portion. In itself, that change should dramatically improve the nutrition of many horses.

Relevance and standardisation of the terms Glycaemic index and Glycaemic response

Patricia A. Harris[1] and Ray J. Geor[2]
[1]Equine Studies Group, WALTHAM Centre for Pet Nutrition, Melton Mowbray, LE14 4RT, United Kingdom
[2]Middleburg Agricultural Research and Extension Center, Virginia Tech, Middleburg, VA 20117, USA

Why important?

Obesity and obesity related metabolic diseases are perhaps today the most important public health concerns facing the developed world. However, advice to reduce total dietary fat in the diet, in an attempt to reduce calorie intake, often leads to reciprocal increases in carbohydrate intake. This in turn may not be without risk to health, as it is appreciated that not all carbohydrates produce the same metabolic effects. In particular they differ in the extent to which they raise blood glucose and insulin concentrations (Aston, 2006). The concept of glycaemic index (GI) was developed approximately 25 years ago (Jenkins *et al.*, 1981) as a means to classify carbohydrate-containing foods based on their blood glucose raising potential. Initially there was little interest or support for the idea, but as more information became available the potential advantages of defining feeds according to their GI became more widely accepted and in 1997 two epidemiological studies were published which suggested that high GI food consumption was associated with an increased risk of type 2 diabetes. Since then there have been numerous studies looking at the role of GI and glycaemic load (GL) in the prevention and management of type 2 diabetes, cardiovascular disease, obesity and other chronic diseases such as cancer (Aston 2006, Feskins and Du, 2006). However, the true impact of changing from a high to a low GI diet remains controversial, with many conflicting studies, in part because study designs are confounded by multiple dietary manipulations, not just a change to a low GI diet (Aston, 2006; Feskins and Du, 2006). In addition, although there are numerous international tables containing published data on the GIs of individual human foods (Foster-Powell *et al.*, 2002) these do not take into account regional or individual differences in foods or food preparation (Feskins and Du, 2006). Despite this obvious limitation, many human studies looking at the influence of GI on health and disease risk use estimated GI values based on such tables and food questionnaires – rather than actual GI data for the foods used (Feskins and Du, 2006).

So why the interest for the horse, a non-ruminant herbivore, suited to consumption of high fibre feeds that are subject to continual microbial fermentation, predominantly within the caecum and colon. It is because of domestication, and an increasing demand for horses to perform at levels that require energy intakes above those able to be provided by their more 'natural' diet of fresh forage, that we commonly include cereal grains and their by-products in horse diets. Under typical management practices horses are fed restricted amounts of forage and provided with one to two large meals a day. This limited feeding time contrasts markedly with the natural situation, where horses may forage for up to 17hrs per day. In addition, modern diets often consist of feedstuffs with greatly reduced water content (e.g. cereal based) and a radically different nutritional profile (high starch, sugar content) compared to the diet that they would be able or would choose to select in the wild. These modern practices have benefits but also potential disadvantages to the horse, both nutritionally and behaviourally, which may have an impact on health and welfare (Davidson and Harris, 2002; Kronfeld and Harris, 2003). In addition, today if horses have access to pasture this tends, especially at certain times of the year, to be much higher quality (i.e. higher protein, sugars and starch content) than the pastures their digestive and metabolic systems evolved to manage. Recent work has suggested that pastures with high levels of starch, sugar and/or fructans can result in marked fluctuations of blood glucose and insulin in a similar way to the feeding of large cereal based meals (McIntosh *et al.*, 2007).

Such marked changes in insulin and glucose concentrations over the day, due to certain pastures or feeding practices, have been linked with abnormalities in growth and an increased risk of developing insulin resistance (which in turn may be linked with an increased risk of laminitis) and/or obesity. It is this link with certain key conditions that has resulted in the increased interest in diets for horses that do not result in marked post-feeding insulin and glucose responses but still enable them to grow, perform and live active healthy lives (Kronfeld *et al.*, 2005; Trieber *et al.*, 2006). It has been, for example, suggested that diets, which produce high glycaemic peaks (and subsequent effects on insulin and other hormones), may increase the risk of DOD (Glade and Belling, 1986; Ralston, 1995 and 1996; Pagan, 2001). Such diets have the potential to establish a feeding fasting cycle, which is a perturbation from hormonal patterns likely seen in animals grazing in the wild. This in turn may adversely influence bone development, as the cyclical changes in glucose and/or insulin may influence chondrocyte maturation via effects on:

- Growth hormone (Freud *et al.*, 1939).
- Thyroxine and triiodothyronine (Glade and Reimers, 1985).
- Insulin-like Growth Factor I (Cymbaluk and Smart, 1993; Henson *et al.*, 1997; Staniar *et al.*, 2001; Staniar *et al.*, 2002; Burk *et al.*, 2003).

As in man, insulin sensitivity is affected by diet in the horse. In humans, studies have shown that even in non-diabetic subjects low GI diets may reduce certain markers of insulin resistance in the short and long term (Feskins and Du, 2006). Studies in the horse have shown lower insulin sensitivity in untrained, sedentary Thoroughbreds (Hoffman *et al.*, 2003) and Standardbreds (Pratt *et al.*, 2006) and a significant reduction in Thoroughbred weanlings (Treiber *et al.* 2005a) when adapted to starch and sugar rich supplementary feeds compared with those adapted to fat and fibre rich feeds. This decrease in insulin sensitivity may be particularly important in animals with pre-existing insulin resistance, which is now recognised as an important predisposing factor for some forms of laminitis (Trieber *et al.*, 2005b, Treiber *et al.*, 2006).

The increasing awareness of the potential for diet composition, particularly the starch and sugar content, to affect the short and long term health of horses, has resulted in an interest in ways to mitigate the metabolic consequences of certain diets. In parallel with the developments in the human field, there has, therefore, been an increase in work examining the effects of diet on glycemic responses in the horse as well as interest from the feed manufacturers in producing diets that moderate these responses. This has led to an increase in the use of terms such as 'low glycaemic'; 'low glycaemic index (GI)' with respect to horse feeds.

Application in man

Definitions

Available carbohydrates

Available carbohydrates (CHO) e.g. simple sugars such as lactose, glucose, sucrose, fructose, (maltodextrins, maltose) and available starch. These can be digested by mammalian enzymes within the small intestine (SI) to hexoses, which can be absorbed from the SI, or if they 'escape' digestion in the SI they can be rapidly fermented (often to lactic acid as well as other short chain fatty acids: SCFA) in the hindgut.

Starch

Starch consists of polymers of glucose, which occur in two forms: amylose and amylopectin. The former is a linear α - (1-4) linked molecule, whereas the latter is a larger molecule, being highly branched containing both α - (1-4) and α - (1-6) linkages. Starch contains varying proportions of amylose and amylopectin, the ratio depending upon the botanical origin of the starch and even within one type of cereal the proportions can vary considerably, for example, corn starch is normally around 74 -76% amylopectin but some varieties (called 'waxy') can have over 99% amylopectin and yet other modified lines can have up to 90% amylose. In addition, to the variation in the ratio of amylose and amylopectin, the molecular weight of starch can vary greatly which may influence digestibility.

Glycaemic response

The incremental area under the plasma glucose vs. time curve elicited by a meal. There are several ways of determining the AUC but the most commonly used is the incremental method which includes the area over the baseline glucose concentration ignoring the area of the curve that falls below the baseline.

Glycaemic index

A classification of the blood glucose raising potential of carbohydrate foods (Granfeldt *et al.*, 2006). Wolever (2006) has defined glycaemic index as the ratio of the incremental area under the glycaemic response curve elicited by the test food (F) providing a fixed amount of carbohydrate and the reference food (R) providing the same amount of carbohydrate when fed on separate occasions to the same subject i.e.:

$$GI = 100 \text{ x } AUC_F / AUC_R$$

Fundamentally foods with a high GI produce a higher peak and greater overall blood glucose response than those with a low GI, (which release glucose into the blood at a slower rate). In man a low GI food has been defined as one with a GI of ≤ 55, whilst a high GI food has a value ≥ 70.

Ideally the value of the AUC_R should be the average of two or three tests of the reference food on separate occasions.

The GI value has typically been taken to reflect the response to 50g of *available* carbohydrate in the test food in comparison with either 50g of glucose or 50g of available carbohydrate from white bread. However, one of the concerns is how to estimate the available carbohydrate load – one definition is the 'total carbohydrate by difference minus dietary fibre (DF) analysed by the AOAC method ' but this method may overestimate the available glycaemic carbohydrate load (and therefore result in smaller portion sizes). An alternative approach is to use total starch minus resistant starch (RS; see Brouns *et al.*, 2005; Granfeldt *et al.*, 2006).

It is important to note that the GI is a biological measurement i.e. it reflects the response in people rather than being an analytical test in the laboratory. This has resulted in a number of criticisms because of the cost and difficulty in determining the GI of a food, the inherent variability of the results and misunderstandings with respect to the results and what they might mean (Wolever, 2006).

Relative glucose or insulin response or relative glycaemic effect (RGE)

The relative glucose response has been used to compare the glucose or insulin response elicited by a portion of food containing 50g *total CHO* relative to 50 or 75g glucose.

Glycaemic load (GL)

This has been defined as the GI x the amount of available CHO the food portion contains i.e., the product of GI and the amount of glycaemic carbohydrates of a serving of food (Brand Miller *et al.*, 2003).

Some authors use a relative GL term (RGL) i.e. GI/100g of carbohydrate (meaning sugars and starches i.e. available CHO).

A low GL diet can therefore be achieved either by reducing the carbohydrate intake or by reducing the GI of the carbohydrates consumed.

Glycaemic Glucose Equivalent (GGE)

This is defined as 'the weight of glucose equivalent in its effect to a given weight of food' and has been suggested by some to be a more accurate predictor of glycaemic response of a complex meal but this has been challenged.

= Food weight X the % CHOAVL x (GI for that food/GI for glucose)

where % CHOAVL is the amount of available CHO contained in 100g food.

Use of these terms

It has been suggested that the term GI should be used when testing foods based on *available CHO* whereas GL, RGE or GGE should be used to classify the glycaemic impact of foods based on *total CHO* or *serving size*.

Factors influencing GI in man

The GI of a particular feed can be affected by a number of factors, many of which are of relevance to the horse, and some of these are outlined below (see also Wolever, 2006; Brouns *et al.*, 2005; Fardet *et al.*, 2006).

Characteristics of the feed
- The type of starch e.g. *more amylopectin results in increased GI; more amylose results in reduced GI.*

- The rate of digestion of carbohydrate within a particular feed e.g. *extruding or puffing starch increases the glycaemic response, parboiling can reduce the GI, the degree of 'destructuring' in the mouth and stomach influences how much of the starch granules remain embedded in the food matrix and therefore their accessibility to the pancreatic amylases; interactions between starch and proteins/lipids and fibres within the feed influence their accessibility to amylase e.g. in white bread the gelatinised starch granules are embedded in a relatively thin protein network; gluten-free bread has a higher GR than normal bread; some grains have stronger starch – protein interactions e.g. hard wheat vs. soft wheat.*
- Rate of gastric emptying – which in turn is influenced by the nature of the food (*e.g. high organic acids reduce gastric emptying*).
- The nature of the monosaccharides absorbed e.g. *in man fructose and galactose do not raise blood glucose significantly.*
- Insulin response elicited by the feed e.g. *the proteins in milk and other dairy products elicit particularly high insulin responses.*
- Effect of dietary fibre – *inclusion of soluble viscous fibre (e.g. guar, B-glucans – provided they have not been depolymerised during the 'cooking process' and form an intrinsic part of the food matrix) may increase the viscosity of the digestive medium so limiting the diffusion and hence the absorption of glucose through epithelial cells and therefore tend to decrease GI.*
- Presence of resistant starch – *more RS means a reduced GI.*
- Presence of certain anti-nutritive factors *e.g. those that inhibit amylase activity reduce GI.*
- Nature of the previous meal e.g. *a low GI barley meal given as the evening meal significantly reduced the GI (as well as the insulin index) of white wheat flour given as the breakfast meal; the response to a morning glucose load was lower if a low GI lentil dinner had been fed, rather than a high GI meal (glucose), the night before.*
- Time of day/exercise level prior to the test/smoking etc.
- The method for determination of AUC.

Can in vitro evaluations replace the GI?

The above highlights the fact that the GI or GR are biological/ physiological measurements, not theoretical values based on the chemical composition of the feed. Currently the results from *in vitro* digestion techniques do not always correlate well with *in vivo* measurements of glycaemic response and therefore do not necessarily provide accurate estimates of the GI (Wolever, 2006). Future improvements may, however, help improve the accuracy of *in vitro* tests. For example, understanding in more detail how the different types of starch affect postprandial metabolism will enable corrections to be made according the relative amounts of predominantly rapidly digestible versus slowly digestible starch in the particular food (Ells *et al.*, 2005). Other studies, in man at least, have suggested that if the sample preparation mimics the effects of chewing then *in vitro* enzymatic procedures can be used to facilitate the ranking of foods with respect to their predicted glycaemic response (Bjorck *et al.*, 1994) even if they do not provide identical GI values.

It is also useful to note that it is very difficult to generalise about the GI values of particular food groups, as for example, the GI for potatoes can vary from 25 – 111, bread from 27 (barley bread with 75% whole grains) to 95 (extremely porous French baguette) and pasta from 27 - 78 depending on variety, processing and cooking (Foster-Powell *et al.*, 2002; Wolever, 2006).

Finally several authors point out that GI is largely irrelevant for foods that contain small amounts of carbohydrate per serving (such as most vegetables; Flight and Clifton, 2006). Brouns *et al.* (2005) point out that the GI should only be applied to foods in which carbohydrate contributes >80% of the energy content.

Variability in individual response

Random variability in glycaemic responses from day to day is reflected by variation in the GI value for a particular food in an individual person (the average coefficient of variation [CV = 100 x standard deviation/ mean] for normal subjects for the AUC is around 25%) - this means that on any one occasion a value could be between + or – 50% of its true average value. Therefore food GI values should be based on the average glycaemic response of a large number of subjects rather than the response within or between a few individuals.

Standardisation

Unfortunately 'standard' methods for determining available CHO and undertaking GI measurements are not applied in man at the moment although it has been recognised that such standardisation would be very useful (see Brouns *et al.*, 2005; Granfeldt *et al.*, 2006).

Application of GI to mixed meals

It is argued by some that the GI of individual foods does not predict the glycaemic impact of mixed meals. One explanation is that 'the confounding effects of fat and protein, along with many other subject- and meal related variables on glycaemic responses overwhelm any differences due to the GI of individual foods, which are unpredictable anyway because of variation in cooking, processing, chewing etc' (Wolever, 2006). However, Wolever has argued that 'every study concluding that GI does not apply in mixed meals suffers from methodological flaws that render that the conclusion invalid' and reports that over 90% of the variation in observed mean glycaemic responses could be explained by the GI and carbohydrate contents of the foods in the meals.

Why not use the insulin response rather than the glycaemic response or index?

Again this issue is hotly disputed in the human literature. It can be argued that insulin, not glucose is the most relevant blood marker of meal response to carbohydrate as raised insulin concentrations have been linked with increased fat storage, insulin resistance, pancreatic B-cell exhaustion and hyperlipidaemia (Wolever, 2006). However, there is some evidence that unlike the GI the relative insulin responses to foods may be significantly different in different groups of people and between certain individuals and may be inversely related to the individual's fasting plasma insulin (Wolever *et al.*, 2004). In one study, of potential relevance to the horse, the responses were compared in four groups of subjects (Lean normal, Obese normal, Impaired glucose tolerance, Type 2 diabetes). The mean relative glucose response was similar in all four groups (Wolever *et al.*, 1998) whereas the insulin responses were significantly different (mean values for the four groups being 59, 50, 69 and 84%). This suggests that the relative insulin response elicited by a food depends on the subject in whom it is tested and therefore might not be a valid property of that food – although it might be an important variable to monitor for the individual.

Situation in the horse

As the digestive processes in the horse are different from those in man it is useful to start by considering starch and sugar digestion in the horse.

The stomach volume of the adult horse is relatively small (between 9 – 15 litres for a 500kg horse). The stomach is relatively inelastic and has a finite capacity and as the rate of gastric emptying is dependent on the square root of the volume, effectively, the larger the meal the more rapid the rate of gastric emptying. There is also a knock-on effect, as the ingesta will then pass more quickly through the small intestine. The stomach is divided into two sections, which have both anatomical and physiological differences. In the cranial non-glandular section (squamous part), bacterial fermentation of the ingested feed starts. This mainly involves lactobacteria, which convert any available simple sugars or starches predominantly to lactic acid. This may be an important point when discussing glycaemic responses in horses and comparing the results with those reported in man and other animals where such extensive gastric fermentation of available CHO does not occur. This microbial activity and degradation is stopped when the gastric contents pass to the fundic gland region and mix with the acid stomach juice containing pepsinogen. The saliva of a horse contains little, if any, amylase and little enzymatic digestion occurs in the stomach of most horses.

Most digestion in the horse, therefore, occurs in the small and large intestine. The basic digestive processes (enzymatic degradation of proteins, fats, starches and sugars) are similar to those of other monogastric animals, and, for example, the end product of starch digestion in the small intestine of the horse is glucose. However, the activity of most of the enzymes in the chyme, in particular amylase, is lower in horses than in other monogastric animals. The horse, therefore, has a limited capacity to digest starch in the small intestine. The exact limit does seem to vary with the individual, and although feeding high starch diets may result in increased levels of amylase, the increases are not marked. The extent of any starch digestion depends in part on the feedstuff with particular respect to the ratio of amylose to amylopectin. This is because the linear nature of amylose enables strong bonds to form aligned chains rendering it relatively resistant to mammalian amylolytic enzymes. By contrast, the highly branched amylopectin molecules cannot be so aligned, and are more easily degraded by mammalian amylases. Differences in starch composition of cereal grains, which have yet to be fully determined, will influence the extent of starch degradation in the small intestine. Thus, oat starch appears to well digested in the foregut of horses (85%), whereas that of unprocessed or rolled barley is poor (21%) (Meyer *et al.*, 1993, Meyer *et al.*, 1995 Kienzle *et al.*, 1998). Thermal treatment of grains to swell and gelatinise the starch improves foregut starch digestibility (Granfeldt *et al.*, 1994; McLean, 2000). However, during cooling after heat treatment, some forms of starch may become resistant to mammalian digestion, due to starch retrogradation to an indigestible, crystalline form (Englyst and Cummings, 1987). Starch not digested in the small intestine, due to its chemical composition or because of physical entrapment within intact cell walls, is termed resistant starch (RS).

Ingestion of excessive levels of starch may exceed the relatively limited amylolytic capacity of the equine foregut; the undigested starch, together with any RS (not fermented in the SI by the microbes present there) passing into the large intestine. The large intestine does not have mucosal enzymes and does not have any significant active transport mechanisms for hexoses and amino acids. Digestion and absorption of residual carbohydrates relies instead on microbial action and absorption of the end products of microbial fermentation. The intensity of these processes depends on the amount and the temporal influx of fermentable material arriving from the small intestine. Large grain meals may therefore overwhelm the digestive capacity of the foregut leading to the rapid fermentation of the grain carbohydrate in the hindgut (producing in particular lactic acid) and a decrease in the pH. A significantly decreased caecal pH may initiate a serious chain of events including a change in the microbacterial flora (excessive growth of those bacteria that can live under such conditions), a degree of lysis of those bacteria which cannot live at such low pHs allowing the release of endotoxins and other factors, damage to the mucosa of the caecum and colon which in turn may allow the increased absorption of endotoxin and various other factors with potential clinical consequences including colic, diarrhoea and laminitis (Kronfeld and Harris 2003; Harris and Arkell, 2006).

The above highlights the fact that the type and amount of starch and sugar containing feed that is ingested should, as in man, influence the blood glucose and insulin responses in the horse. However, there are perhaps three major differences:

- The potential for fermentation of available carbohydrate in the stomach of the horse. The extent of this fermentation will depend on the individual's gastric flora, the nature of the feed, various other management factors as well as gastric emptying rate. The end products of such fermentation are mainly lactic acid plus some short chain fatty acids rather than glucose, effectively reducing the available CHO reaching the small intestine. There may also be consequences for gastric health.
- Greater short term risks associated with high starch/sugar intakes due to the limited amylase activity and the potential for dramatic adverse consequences if large amounts of rapidly fermentable starch/sugar reach the hindgut of the horse. Therefore high GL diets may cause short term problems unrelated to any long term metabolic influence of the feed GI. Therefore, a low glycaemic response *per se* may not be the desired response if it means most of the starch has bypassed the small intestine (e.g. high amylose variety cereals) especially if large amounts are fed.
- In humans, fructose has a smaller influence on serum insulin concentrations than glucose and no influence on plasma glucose levels – in the horse, however, there was no appreciable difference in the insulin or glucose concentrations when fructose, glucose or a mixture was fed to resting animals although there were some minor differences when these were fed in-between two exercise bouts (Bullimore *et al.*, 2000). In another study the effects of higher quantities of fructose and glucose were compared and it was suggested that in resting horses glucose does result in significantly higher glycaemic and insulinaemic responses compared to fructose. Fructose feeding during exercise resulted in lower but still marked glycaemic responses, although there was no effect on the exercise associated insulin response (Vervuert *et al.*, 2004).

So what do we know about responses in the horse. The following provides a non-exhaustive list of some of the studies which have illustrated the effects of some of the factors that can influence the glycaemic response to a feed:

Type of cereal

- Sweet feed, oats and corn produced higher glycaemic indexes than barley, flaked rice bran, loose beet pulp, loose soy hulls and flaked wheat bran (Rodiek and Stull, 2005).
- The AUCs for glucose for Thoroughbreds fed cracked corn, oat groats and rolled barley providing approximately 2g (starch and sugar)/kg BW (Jose-Cunilleras *et al.*, 2004) were not significantly different, although barley was the lowest at 468 ± 42mM/min compared with 519 ± 106 for the corn and 514 ± 43 for the oats. However, this apparent difference was decreased when the glucose AUC was adjusted for actual hydrolysable carbohydrate ingestion. The GI compared with glucose was 63% for the corn and oat groats and 57% for the barley. The authors noted that ingestion of corn resulted in the largest fluctuations in plasma glucose.
- Sweet feeds and whole oats showed the greatest glycaemic response while alfalfa and sweetfeed plus corn oil showed the lowest response in 6 Thoroughbreds when the results of feeding 0.75 kg, 1.5 kg and 2.5 kg of each feed were combined. No significant differences were found in the glucose AUC for whole oats and cracked corn (Pagan *et al.*, 1999a).
- No differences were seen in the glycaemic response in Standardbreds fed whole oats, barley or corn at a moderate starch intake (between 1.2 to 1.5g starch/kg BW: Vervuert *et al.*, 2003, 2004).

Amount of feed

- Increases in GR were found when the feed intake was increased from 0.75 to 1.5 or 2.5 kg per meal for sweet feed, high fibre mix, and sweet feed plus oil. However, this pattern was not seen with whole oats (highest AUC seen with 1.5 kg intake). With cracked corn similar results were seen for the medium and high intakes, both with higher GR than the low intake and for alfalfa the medium intake gave a lower GR than the low intake which in turn was lower than with the high intake (Pagan *et al.*, 1999).

- A lack of a systematic dose response relationship has been seen with oats as there was an appreciable drop in the GI for whole oats fed at 2.5 kg compared to that at 1.5 kg (Pagan *et al.*, 1999).
- Increasing the amount of oat starch from ~1.2 to 2-4 g starch/kg BW resulted in higher glycaemic responses (Vervuert *et al.*, 2003, 2004).
- Feeding the same amount and type of feed either in 2 (~2.6 kg/meal), 3 (~1.75 kg/meal) or 4 (~1.3 kg/meal) meals a day suggested that larger concentrate meals fed in a twice a day schedule would result in a greater plasma glucose response than if the same amount of feed was divided into more than 2 meals. Although there were significant differences with respect to glucose AUC, it could be questioned whether these differences were clinically significant (1,253.7 vs. 1168.7 vs. 1,176 mg/dl; Steeleman *et al.*, 2006).

Effect of processing

- When the glycaemic response of cracked, ground or steam processed corn was compared with that of cracked corn – the highest GI (expressed relative to cracked corn) was for steam- flaked corn fed at 2g/kg BW/meal (Hoelstra *et al.*, 1999).
- No difference was found following mechanical or thermal processing of oats, barley or corn with a moderate starch intake (between 1.2 and 1.5 g starch/kg BW /meal) but a difference was found when the starch intake was increased to 2 g starch/kg BW (Vervuert *et al.*, 2003, 2004).
- In a study looking at the effects of different processing procedures for barley the lowest degree of gelatinisation was found in whole barley (14.9%) and the highest in popped barley (95.6%) however there was no difference in their GR (*note that RS was not determined*). Finely ground barley GR was significantly lower than that of whole barley (Vervuert *et al.*, 2005). However, as the difference in the mean glucose AUC between the feeds was <5% of the highest AUC, effective ranking of the diets was not possible.
- Different glycaemic responses were found when similar starch content was fed as a pellet or a sweet feed (Harbour *et al.*, 2003).
- No difference in postprandial glycaemic response was found in weanling Standardbred horses fed (at 3% in two meals) the same ration either as alfalfa hay cubes and a texturised grain mix (50:50) fed separately or the same mix ground and pelleted (Andrew *et al.*, 2006).

Effect of fibre addition

- No difference was apparently found in the postprandial AUC for glucose between 100% alfalfa, 100% corn or a combined corn/alfalfa diet (Stull and Rodiek, 1988).
- Feeding long hay before or with a sweet feed resulted in a reduction of the glycaemic response (Pagan and Harris, 1999).
- Feeding up to 35% of short chop lucerne to oats or a sweet feed did not influence the glucose AUC (Harris *et al.*, 2005a,b).

Note that the glycaemic response within a particular forage type can vary considerably especially as in some legumes such as alfalfa the amount of starch can be very variable (see Kronfeld et al.*, 2004).*

Relationship between insulin and glucose responses

- Similar rankings were found in adult QH fed mixed diets with whole barley and oats; sugarbeet pulp (soaked), grass meal and soybean oil; rice bran and grass meal; rice bran, grass meal, sugarbeet meal and soybean oil (Zeyner *et al.*, 2006) but the actual values between the II and GI (related to the cereal meal) were not very close and the correlations between glucose and insulin were between 0.364 – 0.570.
- Different rankings were found between the GI and Insulin Index (II) (Vervuert *et al.*, 2003, 2004).
- In one study although the rankings would have been similar if the GI or II had been used (i.e. corn, oats, barley) using GI there was very little difference between the corn and oats with the barley only slightly lower than the others and with no statistical difference between them – for the II although there was no significant difference between the feeds the barley had a noticeably lower insulin AUC (P = 0.06) (Jose-cunilleras *et al.*, 2004).

Effect of physiological state

- Response to a glucose load differed according to reproductive state (Hoffman *et al.*, 2003, Williams *et al.*, 2001; Cubitt, 2007).
- Response to a glucose load different between healthy ponies, ponies which had previously suffered from laminitis and Standardbred horses (Jeffcott *et al.*, 1986).

Rodiek and Stull (2007) point out that in their study, which examined the glycaemic responses to 10 common equine feeds, variation in the actual calories of each feed offered (4- 6 Mcal DE), time to complete consumption (15- 300 min), feed refusal (14, 31, 37% for soyhulls, rice bran and beet pulp), feed processing as well as content and form of the starch and sugar likely contributed to the variation in the data. It is also important to note that if we are to use glucose as the reference material many factors influence the response to a glucose challenge including prior diet (Jacobs and Bolton, 1982).

This all suggests that many of the same factors identified as influencing the GI in man will also have an effect in the horse, including the type and amount of hydrolysable carbohydrate fed as well as the individual animal. Obviously due to the inherent risks of large amounts of starch or sugar reaching the hindgut and being rapidly fermented, in the future we need also to consider ways to check that a low glycaemic response has not resulted in disturbances in hind gut flora. This perhaps suggests that we should be working towards considering adapting the idea of determining the fore and hind gut glycaemic load of a feed! However, as an initial step, as has been suggested by previous authors on this topic (Vervuert and Coenen, 2006) we need to firstly encourage feed manufacturers to test diets that they claim have a low GI or GR to confirm that this is true at least in a target group of animals. Second, we must develop a standardised methodology for evaluation of GR in the horse. Initially we need to decide whether we want to continue to use the term GI and test foods according to this protocol (or an amended upscaled protocol) i.e. use the level of available carbohydrate within the food to dictate how much food we should feed – OR use and adapt the GL/RGE or GGE approach.

In doing this we might need to take into consideration whether we should:
- fast overnight (or for how long) or not before and during the test;
- restrict exercise intensity the day (?) before any test;
- include the response to a set amount of glucose, in each study so that the relative GR can be determined and/or agree an additional reference feed e.g. dehulled oats?
- undertake a minimum of 2-3 tests per animal per diet;
- note BCS /breed/age/reproductive state/stage of oestrus cycle to allow comparison or potentially help explain differences;

- determine as a minimum basal blood insulin concentrations and ideally level of insulin resistance;
- determine how we determine and what should be our standard amount of 'carbohydrate' e.g. starch or starch and sugar that is fed per meal and then use this;
- allow a set time for food ingestion and note the duration of feed intake and determine actual starch/starch & sugar ingested in that time;
- agree how to determine the AUC;
- characterise the feeds in a standardised way:
  — degree of starch gelatinisation and amount of RS;
  — percentage of amylose;
  — microscopically: Starch interactions with other food components;
  — food matrix porosity;
  — (ideally determine the food particle size that would be obtained during the upper part of the digestive process and therefore be presented to the SI).

In particular:
- How do we define available carbohydrate in the horse – (i.e. starch, sugar (excluding fructans) minus RS?).
- How much should we feed (i.e. base on the 50-g for humans and scale up to perhaps 350g for a 500 kg horse?).

Conclusions

There is mounting evidence that there may be health benefits associated with the provision of low GI foods to both man and the horse. These health benefits tend to be linked with their slow rate of digestion and reduced postprandial insulin responses, coupled with factors such as reduced fluctuations in blood glucose, or differences in gut hormone responses. Part of the problem seems to be how to translate – accurately – the science into practical advice for the individual. For humans does it mean avoid refined foods, eat more fruits and vegetables instead of concentrated juice and have more pasta and less potato or, rather, should we be saying 'choose high carbohydrate foods (breads, cereals, grains, fruit, dairy products) with a lower GI' (Wolever, 2006). Certainly in the horse the provision of diets with low GI or GR claims but no evidence of this will confuse the situation. Currently we do not have enough information to be able to formulate diets that we can confidently say will result in a suitable GR without these diets being evaluated in the horse itself. Ideally we should also agree on a suitable methodology for this testing. In the meantime it would seem prudent to feed certain groups of horses (e.g. growing horse, senior horses, those prone to laminitis) diets which are have been shown to produce a low (or moderate) glycaemic response or at least avoid feeding diets that are likely to produce a high GR. However, much more work is needed in this whole area.

References

Andrew, J.E., Kline, K.H. and Smoth, J.L., 2006. Effects of feed form on growth and blood glucose in weanling horses. J Equine Veterinary Science 26 349 355.

Aston, L.M., 2006. glycaemic index and metabolic disease risk. Proceedings of the Nutrition Society 65 125 – 134.

Bjorck, I.. Grnfeldt, Y., Liljeberg, H., Tovar, J. and Asp, N.-G., 1994. Food properties affecting the digestion and absorption of carbohydrates. Am J Clin Nutr 59 (suppl) 699S - 705S.

Brand-Miller, J., Thomas, M., Swan, V., Ahmad, Z., Petcock, P. and Claggier, S., 2003. Physiological validation of the concept of glycemic load in lean young adults. J Nutr 133 2728 – 2732.

Brouns, F., Bjock, I., Frayn, K.N., Gibbs, A.L., Lang, V., Slama, G. and Wolevar, T.M.S., 2005. Glycaemic index methodology. Nutr Res Rev 145 – 171.

Burk, R., Staniar, W.B., Kronfeld, D.S. Akers, R.M. and Harris, P.A., 2003. Insulin-like growth factor binding proteins fluctuate with age in growing Thoroughbred foals. Proc Equine Nutr. Phys. Soc. 18 84.

Cubitt, T.A., George, C.A., Staniar, W.B., Harris, P.A. and Geor, R.J., 2007. Glucose and insulin dynamics during the estrous cycle of thoroughbred mares . In Proceedings of the 20th ESS Symposium Maryland p 44 – 45.

Cymbaluk, N.F. and Smart, M.E., 1993. A review of possible metabolic relationships of copper to equine bone disease. Equine Vet J suppl 16 19 – 26.

Davidson, N. and Harris, P., 2002. Nutrition and Welfare. In The welfare of horses Waran N (ed) Kluwer Academic publishers Netherlands 45 – 76.

Els, L.J., Seal, C.J., Keelitz, B., Bal, W. and Mathers, J.C., 2005. Postprandal glycaemic, lipaemic and haemostatic responses to ingestion of rapidly and slowly digested starches in healthy young women. British Journal of Nutrition 94 948 – 955.

Fardet, A., Leenhardt, F., Lioger, D., Scalbert, A. and Remesy, C., 2006. Parameters controlling the glycaemic response to breads. Nutrition research reviews 19 18 – 25.

Feskens, E.J.M. and Du, H., 2006. Dietary glycaemic index from an epidemiological point of view. International journal of obesity 30 S66 – S71.

Flight, I. and Clifton, P., 2006. Cereal grains and legumes in the prevention of coronary heart disease and stroke: a review of the literature European Journal of Clinical Nutrition 60 1145 – 1159.

Freud, J., Levie, L.H. and Kroon, D.B., 1939. Observations on growth (chondrotrophic) hormone and localization of its point of attack. J. Endocrinol. 1: 56 – 64.

Foster-Powell, K., Holt, S.H. and BrandMiller, J.C., 2002. International table of Glycemic index and glycemic load values A J Clin Nutr. 76 5 – 56.

Glade, M.J. and Belling, T.H., 1986. A dietary etiology for osteochondroitic cartilage, Journal of Equine Veterinary Science 6 175 – 187.

Granfeldt, Y. Liljeberg, H. Drews, A., Newman, R. and Bjorck, I., 1994. Glucose and insulin responses to barley products: influence of food structure and amylose amylopectin ratio Am J Clin Nutr 59 1075 – 1082.

Granfeldt, Y., Wu, X. and Bjorck, I., 2006. Determination of glycaemic index some methodological aspects related to the analysis of carbohydrate load and characteristics of the previous evening meal European Journal of Clinical Nutritin 60 104 – 112.

Harbour, L.E., Lawrence, L.M., Hayes, S.H. Stine, C.J. and Powell, D.M., 2003. Cncentrate compostion form and glycaemic response in horses. Proc 18 Equine Nutr Phys Symp 329 – 220.

Harris, P.A., Sillence, M., Inglis, R., Siever Kelly, C., Friend, M., Munn, K. and Davidson, H., 2005a. Effect of short (< 2cm) lucerne chaff addition on the intake rate and glycaemic response of a sweet feed. Pferdeheilkunde 21 88 – 89.

Harris, P.A., Sillence, M. Inglis, R., Siever-Kelly, C., Friend, M., Munn, K. and Davidson, H., 2005b. Effect of short (< 2cm) lucerne chaff addition on the intake rate and glycaemic response to an oat meal Proceedings of the 19th Equine Science Society symposium. Tuscon p151 – 152.

Hoffman, R.M., Kronfeld, D.S., Cooper, W.L. and Harris, P.A., 2003. Glucose clearance in grazing mares is affected by diet pregnancy and lactation. J of Anim. Sci 81 1764 – 1771.

Hoffman, R.M., Boston, R.C., Stefanovski, D., Kronfeld, D.S. and Harris, P.A., 2003. Obesity and diet affect glucose dynamics and insulin sensitivity in Thoroughbred gelding. J Anim Sci 81 2333 – 2342.

Hoekstra, K.E., Newman, K., Kennedy, M.A.P. and Pagan, J.D., 1999. Effect of corn processing on glycaemic responses in horses. Proc 16th equine Nutr Phys Symp 144 – 148.

Jacobs, K.A. and Bolton, J.R., 1982. Effect of diet on the oral glucose tolerance test in the horse. JAVMA 180 (8) 884 – 886.

Jeffcott, L.B., Field, J.R., McLean, J.G. and O'Dea, K., 1986. glucose tolerance and insulin sensitivity in ponies and standardbred horses Equine Vet J 18 97 - 101.

Jenkins, D. Wolever, R., Taylor, R., Barker, H. and Fielden, H., 1981. Glycaemic index of foods: a physiological basis for carbohydrate exchange Am J Cin Nutr 34 362 – 366.

Jose- Cunilleras, E., Taylor, L.E. and Hinchcliffe, K.W., 2004. Glycaemic index of cracked corn, oat groats and rolled barley in horses. J Anim Sci 82 2623 – 2629.

Kienzle, E., Pohlenz, J. and Radicke, S., 1998. Microscopy of starch digestion in the horse 80 213 – 216.

Kronfeld, D. and Harris, P., 2003. Equine grain-associated disorders. Compend Contin Educ Pract Vet.;25:974 – 83.

Kronfeld, D., Rodiek, A. and Stull, C., 2004. Glycemic indices, glycemic loads and glycemic dietetics. J Equine Vet Sci 24 (9) 399 – 404.

Kronfeld, D., Treiber, K., Hess, T. and Boston, R., 2005. Insulin resistance in the horse: definition, detection and dietetics. J Anim Sci 83 E22 – E33.

McIntosh, B., Kronfeld, D., Geor, R., Staniar, W., Longland, A., Gay, L., Ward, D. and Harris, P., 2007. Circadian and seasonal fluctuations of glucose and insulin concentrations in horses. In Proceedings of the Equine Science Society p 100 – 101.

Meyer, H., Radicke, S., Kienzle, E., Wilke, S., Kleffken, D. and Illenseer, M. 1995. Investigations on preileal digestion of starch grain, potato and manioc in horses. J vet Med A 42 371 – 381.

Pagan, J.D., Harris, P.A., Kennedy, M.A.P., Davidson, N. and Hoekstra, K.E., 1999a. Feed type and intake affects glycemic response in thoroughbred horses. Equine Nutrition and Physiology Symposium Proceedings 16 149 – 150.

Pagan, J.D. and Harris, P.A., 1999b. The effects of timing and amount of forage and grain on exercise response in Thoroughbred Horses. Equine Vet J. Suppl 30 Jeffcott L (ed): 451 – 458.

Pagan, J.D., 2001. The relationship between glycaemic response and the incidence of OCD in thoroughbred weanlings a field Study. In Proc of the 47th Annual Conf of AAEP.

Pratt, S.E., Geor, R.J. and McCutcheon, L.J., 2006. Effects of dietary energy sources and physical conditioning on insulin sensitivity and glucose tolerance in standardbred horses. Equine Vet J Suppl 36 579 – 584.

Ralston, S.L., Nockels, C.F. and Squires, E.L., 1988. Diggerences in diagnostic test results and hematologic data between aged and young horses Am J Vet Res 49 1387 –1391.

Ralston, S.L., 1995. Postprandial hyperglycemica/hyperinsulinemia in young horses with osteochondritis dissecans lesions. J. Anim. Sci. 73 184 (Abstract).

Ralston, S.L., 1996. Hyperglycemia/hyperinsulinemia after feeding a meal of grain to young horses with osteochondritis dissecans lesions, Pferdeheilkunde 12 320 – 322.

Rodiek, A.V. and Stull, C.L., 2007. Glycaemic index of ten common horse feeds. J of equine Vet Sci 27 (6) 205 - 211.

Staniar, W.B., Kronfeld, D.S., Akers, R.M., Hoffman, R.M., Williams, C.A. and Harris, P.A., 2001. Dietary fat and fiber influence plasma insulin-like growth factor-I: an endocrine link between diet and osteochondrosis. Proc. Am. Acad. Vet. Nutr. July 14,2001, 13 – 16.

Staniar, W.B., Kronfeld, D.S., Akers, R.M., Burk, J.R. and Harris, P.A., 2002. Feeding-fasting cycle in meal fed yearling horses. J. Anim. Sci. 80 (Suppl 1):156.

Steelman, S.M., Michael-Eller, E.M., Gibbs, P.G. and Potter, G.D., 2006. Meal size and feeding frequency influence serum leptin concentrations in yearling horses J Anim Sci 84 2391 – 2398.

Stull, C.L. and Rodiek, A.V., 1988. Responses of blood glucose, insulin and cortisol concentrations to common equine diets J Nutr 206 –213.

Treiber, K., Boston, R., Kronfeld, D., Staniar, W. and Harris, P., 2005a. Insulin resistance and compensation in Thoroughbred weanlings adapted to high-glycaemic meals. J Nutr.;83:2357 – 64.

Treiber, K.H., Hess, T.M., Kronfeld, D.S., Boston, R.C., Geor, R. and Harris, P., 2005b. Insulin resistance and compensation in laminitis-predisposed ponies characterized by the minimal model. Pferdeheilkunde 21 91 – 92.

Trieber, K.H., Kronfeld, D.S. and Geor, R.J., 2006. Insulin resistance in equids –possible role in laminitis J Nutr 136 2094S – 2098S.

Vervuert, I., Coenen, M. and Bothe, C., 2003. Effects of oat processing on the glycaemic and insulin responses in horses J Anim Physiol 87 96 – 104.

Vervuert, I., Coenen, M. and Bothe, C., 2004. Effects of corn processing on the glycaemic and insulinaemic responses in horses. J anim Physiiol 88 348 – 355.

Vervuert, I., Coenen, M. and Bothe, C., 2005. Effects of mechanical or thermal barley processing on glucose and insulin profiles in horses. Proceedings of the 9th ESVCN congress Grugliasco p 118.

Vervuert, I. and Coenen, M. 2005. Glycaemic index of feeds for horses Proceedings Equine Nutrition conference Hannover Pferdeheilkunde 21 79 – 82.

Williams, C.A., Kronfeld, D.S., Stanier, W.B. and Harris, P., 2001. Plasma glucose and insulin responses of Thoroughbred mares fed a meal high in starch and sugar or fat and fiber J Anim Sci 79 2196 – 2201.

Wolever, T.M.S., Chiasson, J.L., Hunt, J.A., Palmason, C., Ross, S.A. and Ryan, E.A., 1998. Similarity of relative glycaemic but not relative insulinaemic responses in normal IGT and diabetic subjects. Nutr Res 18 1667 –1676.

Wolever, T.M.S., Campbell, J.E., Geleva, D. and Anderson, G.H., 2004. High fibre cereal resuces postprandial insuli responses in hyperinsulinaemic but not normoinsulinemic subjects Diabetes care 27 1281 – 1285.

Wolever, T.M.S., 2006. Physiological mechanisms and observed health impacts related to the glycaemic index: some observations. International Journal of Obesity 30 S72 – S78.

Zeyner, A., Hoomeister, C., Einspanier Gottschalk, J.A., Lengwenat, O. and Illies, M., 2006. Glycaemic and insulinaemic responses of Quarter horses to concentrates high in fat and low in soluble carbohydrates Proceedings of the 7th ICEEP conference France, Equine Vet J Suppl 36 643 – 647.

The role of nutrition in colic

Andy E. Durham
The Liphook Equine Hospital, Liphook, Hampshire, GU30 7JGH, United Kingdom; andy@TheLEH.co.uk

Colic in the horse is a condition familiar to all equine practitioners and a problem that many horses experience. Colic has enormous negative consequences in terms of welfare of the afflicted individuals, loss of training and competition days, treatment costs and, sometimes, death of the horse (Tinker *et al.*, 1997a; Traub-Dargatz *et al.*, 2001). Most studies have found the incidence of colic to be approximately 5 cases per 100 horses per year (Archer and Proudman, 2006), although some have found a much greater incidence on certain premises that might then benefit from the application of preventative epidemiologic knowledge (Uhlinger, 1992; Tinker *et al.*, 1997a). Many potential risk factors for colic have been investigated including breed, use, gender, age, anthelmintic use, season, weather, transport, diet, general health care, geographic location, stereotypic behaviour and medical history that have enabled clinicians to offer good evidence-based advice to reduce the incidence of colic (Morris *et al.*, 1989; Cohen *et al.*, 1999; Hillyer *et al.*, 2001, 2002; Archer *et al.*, 2006; Archer and Proudman 2006).

Several studies have confirmed that the risk of colic significantly increases with higher levels of cereal or concentrate feeding (Tinker *et al.*, 1997b; Hudson *et al.*, 2001). Tinker *et al.* (1997b) found horses consuming moderate quantities (2.5-5 kg/day) of concentrate feed to be at an almost 5 times increased risk of colic, and those consuming large quantities (>5 kg/day) to be at a greater than 6 times increased risk in comparison to horses on pasture receiving no concentrate. Cohen *et al.* (2006) found that horses with duodenitis-proximal jejunitis were fed significantly more concentrates than horses with colic due to other causes or lame horses. Some studies have suggested pelleted feeds pose the greatest risk of colic (Morris *et al.*, 1989; Tinker *et al.*, 1997b) but this has not been found by others (Cohen *et al.*, 1999; Little and Blikslager, 2002). In addition to potential adverse effects of concentrate feed *per se*, it is possible that the association between high levels of concentrate feeding and colic could be partly explained by other non-dietary factors associated with high concentrate feeding that might also predispose to colic such as breed (e.g. Thoroughbred [Hudson *et al.*, 2001]) and higher levels of physical activity (Cohen *et al*, 1995, 1999; Kaneene *et al.*, 1997).

The equine hind gut contains a considerable population of fermentative bacteria utilising protein, fibre (e.g. cellulose, hemicellulose [xylan]) and non-fibre carbohydrate (e.g. starches, fructans) arriving in the caecum. Bacterial populations are directly influenced by the quality and quantity of ingesta and indirectly via pH changes following bacterial fermentation of available substrates. Pre-caecal digestibility of dietary starch should ideally be complete and starch arriving in the hindgut is more likely to be associated with adverse health effects than beneficial nutritional gains. The extent of starch digestion in the equine small intestine is poor in comparison with other monogastric species and also varies with different feed types (Cuddeford, 2000; Hintz, 2000; Harris and Arkell, 2005). A high proportion of starch in oats is effectively digested by the equine small intestine whereas the digestibility of starch in maize and barley tends to be low unless they are processed (Hintz, 2000; Hussein *et al.*, 2004).

Starches fed in excess of the limited digestive capacity of the equine small intestine will arrive in the caecum and colon and have adverse consequences that have been the subject of many studies (Goodson *et al.*, 1988; Clarke *et al.*, 1990; Potter *et al.*, 1992; de Fombelle *et al.*, 2001; Drogoul *et al.*, 2001; Julliand *et al.*, 2001; Hussein *et al.*, 2004; Lopes *et al.*, 2004). Additionally, gastric and small intestinal transit may be hastened by the more voluminous chyme associated with high starch feeds, further limiting pre-caecal digestibility and increasing hindgut delivery of non-structural carbohydrate (Clarke *et al.*, 1990; Drogoul *et al.*, 2001; Metayer *et al.*, 2004). Some evidence suggests that starch not degraded in the small intestine may also flow quickly through the caecum and lead to greater fermentative derangements in the ventral colon than in the caecum (Julliand *et al.*, 2001). Arrival in the hindgut of such a readily fermentable carbohydrate source initially increases the rate of bacterial multiplication and results in increased prevalence of acidophilic, starch fermenting bacteria such as *Streptococci* and *Lactobacilli* generating lactate and a fall in pH from the normal 6.7-7.0 to as low as 6.0. This acidification has further effects on bacterial populations including a reduction in fibre-fermenting species. Consequently additional alterations in VFA profiles occur including reduced acetate, increased proprionate and further increased lactate and decreased pH. Potential functional consequences of this series of events are decreased fibre digestibility, mucosal barrier dysfunction, endotoxaemia, gas/frothy distension, dysmotility and displacement/ volvulus.

In addition to incomplete pre-caecal starch digestion, increased grain feeding may have additional adverse effects on the hindgut related to postprandial systemic hypovolaemia and dehydration subsequent to increased upper alimentary secretions, a phenomenon that is much more marked in 'greedy eaters' (Clarke *et al.,* 1988, 1990; Houpt *et al.,* 1988). However, this seems generally to be a short-lived (<3hours) phenomenon (as long as water is freely available) and is considered unlikely to directly promote impaction colics in most horses (Lopes *et al.,* 2004).

Starch fed to horses in excess of 2.5 g/kg bodyweight in one meal will result in more than half of the ingested starch reaching the hindgut (Potter *et al.,* 1992). Consequently it is often suggested that horses should eat no more than 2 g/kg bodyweight of non-structural carbohydrate per meal (Cuddeford, 2000; Hussein *et al.,* 2004; Geor, 2005; Harris and Arkell, 2005). However, even at far lower levels than this (e.g. 1 g/kg bodyweight per meal) there may still be 20% of ingested starch reaching the caecum (Potter *et al.,* 1992) especially if inadequately processed barley or wheat products are fed.

Although it seems intuitively correct to divide concentrate meals into as many small feeds as possible over the course of the day, there is no clear evidence to support the overall benefits of such a strategy. Although there is evidence of modest nutritional benefit when feeding frequency is increased, the practice may be associated with increased stereotypic behaviours in the fed horses and in-contacts (Houpt *et al.,* 1988; Clarke *et al.,* 1990; Cooper *et al.,* 2005; Van Weyenberg *et al.,* 2007). Nevertheless, as a general strategy to improve gastrointestinal health the argument for 'little and often' feeding is compelling (Harris, 2007). Small and frequent starch boluses are less likely to reach the hindgut than large and infrequent boluses with subsequent benefits to the stability of bacterial populations. Concentrate feeds should therefore be divided into as many small feeds as practical and dietary starch intake preferably limited to 1 g/kg (or maximum of 2 g/kg) bodyweight per meal (many cereals contain approximately 50% starch).

Oats facilitate pre-caecal starch digestion and are the preferred cereal for horses, however cooked processing of barley and wheat significantly improve pre-caecal starch digestibility and make extruded, micronised, steam flaked or popped products acceptable (Cuddeford, 2000). For horses requiring a diet of moderate energy density, soluble fibre can also be provided by feeds such as sugar beet pulp that contains significantly more fibre and less starch than cereals despite similar digestible energy content. Relatively high fat diets have become popular in many equine management systems in order to permit the feeding of energy dense diets without the undesirable consequences of non-structural carbohydrate on the hindgut as described above. However, vegetable oils have also been found to significantly reduce the cellulolytic capacity of the equine hindgut, presumably as a consequence of bactericidal fatty acids produced by fermentation of oil arriving in the hindgut (Jansen *et al.,* 2002). Although Cargile *et al.* (2004) found that oil feeding was associated with decreased gastric acid synthesis and increased gastric PGE_2, another study did not find that oil feeding reduced either gastric ulcer number or severity (Frank *et al.,* 2005).

Probiotics are popular amongst horse owners, perhaps as a result of good marketing and the assumption that benefits shown in people will also occur in horses. Until recently there had been minimal evidence-based study of probiotics in horses. Clearly appropriate micro-organisms should be shown to be able to colonise the equine gastrointestinal tract following oral dosing, not induce any adverse effects and to have a demonstrable benefit on gastrointestinal function under normal and/or disease conditions. These criteria were at least partly fulfilled in a study of the ability of *Saccharomyces cerevisiae* to protect hindgut disturbance as a result of starch overload (Medina *et al.,* 2002). However, Weese (2002) raised doubts that some products may not even contain the microbial organisms claimed by the product manufacturer. Further studies by Weese *et al.* (2004) attempted to define potentially beneficial equine-derived bacteria that were able to survive transit through the equine alimentary tract. Unfortunately, subsequent trial of the single selected bacterial species (*Lactobacillus pentosus*) in foals was associated with adverse gastrointestinal signs (Weese and Rousseau, 2005) and the usefulness and safety of bacterial probiotic products is still questionable.

Although pre-caecal digestion is often regarded as synonymous with degradation by endogenous (pancreatic/intestinal) enzymes, some bacterial fermentation will also occur in the stomach and small bowel of the horse (Mackie and Wilkins, 1988). Volatile fatty acids (VFAs) produced by gastric bacterial fermentation have been implicated in the pathogenesis of equine gastric squamous mucosal ulceration (Nadeau *et al.*, 2000, 2003). VFAs within the very low pH of the equine stomach will be largely protonated, lipophilic and able to diffuse into gastric squamous mucosal cells (gastric glandular cells are better protected from acid exposure). The dissociation of the VFA molecules within the higher cytosolic pH may lead to intracellular acidification and cell death. A protective effect of alfalfa feeding (in comparison to hay) has been attributed to buffering or possibly higher calcium content of gastric juice (Nadeau *et al.*, 2000). Predisposing factors for gastric ulcers such as high cereal diets and intermittent feeding (Bell *et al.*, 2007) are likely to promote gastric VFA concentration and lower gastric pH promoting ulcerogenesis as described above. Clearly concentrate feeding cannot be considered in isolation from strategic forage intake and continual or frequent forage access will increase saliva production and buffer gastric acidity. Furthermore, the constant presence of indigestible long fibre in the stomach will reduce acid exposure of the gastric mucosae.

Pasture turnout has generally been found to be associated with a reduced risk of colic in comparison to stabled or partly stabled horses. Hudson *et al.* (2001) found that fully stabled horses were twice as likely to have colic as those at pasture full time and even stronger associations between colic and stabling have been found for simple colon obstructions (Dabareiner and White, 1995; Hillyer *et al.*, 2002). Horses turned out to pasture also appear to carry the lowest risk of gastric ulceration (Bell *et al.*, 2007). Possible confounding factors should be considered however, as the association between stabling and colic may be partly explained by more close observation of stabled horses, short term and seasonal weather patterns associated with increased stabling, or non-dietary effects of confinement such as lack of exercise, psychological stress and stereotypic behaviour (Cohen *et al.*, 1999;

Hillyer *et al.*, 2002; Archer *et al.*, 2006). Furthermore, other studies have indicated that the incidence of certain colic subtypes including sand impactions, grass sickness and duodenitis-proximal jejunitis are associated with *increased* grazing activity (Ragle *et al.*, 1989; McCarthy *et al.*, 2001; Cohen *et al.*, 2006).

Fructans, pectins and β glucans contained in herbage are indigestible and most will reach the hindgut where fermentation will tend to reduce pH as previously discussed for starches. Clearly soluble fibre is a normal nutrient for horses and only excessive quantities or changes in quantity are likely to alter the hindgut bacterial *status quo*. Fructan content of grasses and herbage can, however, be markedly variable between and within grass species (Longland and Cairns, 2000) creating the possibility for hindgut disturbance when grazing is changed. Both temperature and light exposure can also have marked and acute effects on fructan content of grass with potential acute destabilisation of intestinal microflora (Longland and Byrd, 2006). The observed associations between weather, season and change of fields with cases of grass sickness may relate to altered enteric toxicoinfectious bacterial populations (Hunter *et al.*, 1999). Given the generally observed reduction in colic risk in association with pasture turnout, this should be encouraged under most circumstances for at least a few hours daily. However, it is unusual for nutritional quality of grazing or specific quantitative intake to be known for grazing horses. As a high level of fructan ingestion is a potential cause of hindgut disturbance in turned-out horses, known risk factors for higher fructan levels should be considered as part of horse and pasture management such as grass species (e.g. Timothy lower than Perennial ryegrass), timing of turnout (e.g. lower fructan at night than in the daytime; higher fructan in Spring and Autumn; higher fructan on bright cool days) and pasture treatment (e.g. regular cutting increases fructan content) (Longland and Cairns, 2000).

Little information is available on which to base specific recommendations for fibre feeding in horses (Cuddeford, 1999). Hintz (2000) suggested that a minimum of 24% neutral detergent fibre (NDF) or 14% acid detergent fibre (ADF) might be adequate for gastrointestinal health but higher levels were desirable. However, some hays very high in indigestible ADF may promote impaction colics (Cohen and Peloso, 1996; Hudson *et al.*, 2001; Little and Blikslager, 2002). More nutrient dense hays such as alfalfa may increase the risk of colic due to small colon obstructions and enterolithiasis (Morris *et al.*, 1989; Cohen *et al.*, 2000; Hassel *et al.*, 2004) and decrease the risk of colic due to small intestinal strangulations and gastric ulceration (Morris *et al.*, 1989; Nadeau *et al.*, 2000). In practice, undetermined and variable nutritional quality of forage can confound a logical approach to dietary design although ranges of proprietary haylages often have informative nutritional declarations. A minimum of 1-1.5% bodyweight of forage (as dry matter) is usually recommended which is likely to ensure at least 40% NDF and 20% ADF in the overall diet. Forage should ideally be available at all times – with any quantitative restriction perhaps being enforced by narrow-weave haynets, double haynets or 'haybags'.

Given the likely influence of bad dentition on digestion and transit of ingested feed it is of interest that many studies have found no association between dental care and colic (Cohen *et al.*, 1995; Reeves *et al.*, 1996) although a possible protective effect of regular dental care was indicated by Hillyer *et al.* (2002). Furthermore, other studies have found no significant improvement in weight gain, body condition score, digestibility of various nutrients or faecal particle size in horses that had teeth floated in comparison to a control group (Ralston *et al.*, 2001; Carmalt *et al.*, 2004).

Although diet quality and quantity *per se* clearly have a marked influence on the incidence of colic, probably the strongest association between nutrition and colic relates to the increased risk of colic within 1 to 2 weeks following a change in the diet (Cohen *et al.*, 1995; Cohen and Peloso, 1996; Mehdi and Mohammed, 2006) perhaps as a result of abrupt alteration in microflora populations. De Fombelle *et al.* (2001) found that changing diet from hay only to 70% hay and 30% rolled barley lead to significant alterations in hindgut bacterial flora and VFA concentrations with the potential for provoking colic. Additionally, dietary change is a recognised risk factor for *Salmonella* shedding in hospitalised horses further implying disturbance of hindgut microflora (Traub-Dargatz *et al.*, 1990). Thus a vital principle of equine dietary management is to introduce any changes in a gradual fashion with mixing of the 'old' and 'new' diet to phase the dietary change over no less than 2 weeks. The risks of dietary changes apply to both concentrate and forage feeding and a recent change in pasture represents one of the greatest risk factors for grass sickness (Wood *et al.*, 1998). Events such as changing the batch or type of hay, changing the quantity or frequency of feeding and erring from usual feeding times significantly increase the risk of colic (Tinker *et al.*, 1997b; Cohen *et al.*, 1999; Hudson *et al.*, 2001). It has generally been found that the risks associated with a recent change in hay or forage are more marked than the risks associated with a recent change in grain or concentrate fed (Cohen *et al.*, 1999; Hudson *et al.*, 2001; Hillyer *et al.*, 2002). Hillyer *et al.* (2002) found that the risks of simple colon obstruction and distension were greatest within 7 days of a dietary change (forage change OR = 22.00; concentrate change OR = 12.03) but were still significantly increased between 8-14 days of the change (forage change OR = 4.88; concentrate change OR = 3.01). Dietary changes between 15 and 28 days previously were not significantly associated with colic.

Intuitively, it seems correct that the health of the equine hindgut may benefit from stability in its fermentative bacterial population. The probable imbalancing effect on bacterial populations resulting from an abrupt dietary change might actually be mimicked over a shorter and cyclical time scale in horses fed a concentrate feed twice daily in a routine manner (Harris and Arkell, 2005). Short pre-caecal transit times, the arrival of starch in the caecum and prompt acidification associated with cereal meals is likely to lead to significant intra-day variability in hindgut pH and bacterial populations and perhaps promotion of digestive problems.

Seasonal effects on colic incidence have been demonstrated for a large heterogeneous group of colic cases, and more notably for certain colic subtypes including grass sickness (May peak), epiploic foramen entrapment (December-January peak), large colon impaction (December-January peak) and large colon displacement/torsion (March and October peaks) (Archer *et al.*, 2006). Changes in weather have been found to increase the risk of colic in some (Cohen *et al.*, 1999), but not all (Proudman, 1991) studies. Most notably, equine grass sickness has established seasonal and possible short term climatic risk factors (McCarthy *et al.*, 2001). Putative causal links between seasonal or weather changes and colic could involve numerous possibilities such as physical activity, foaling, stabling and stereotypic behaviours although dietary factors may well be of great importance also.

Perhaps the most simple and direct associations between dietary intake and colic are occasional reports of intestinal obstruction by ingestion of inappropriate materials such as fencing material, wood, persimmon and a feedblock (Getty *et al.*, 1976; Green and Tong, 1988; Kellam *et al.*, 2000; Mair, 2002). Such cases are rare and generally seen in young horses.

As the gastrointestinal tract experiences the most intimate contact with ingested nutrients, it is hardly surprising that many dietary factors have been shown to represent significant risk factors for colic in the horse. Equine digestive behaviour, anatomy and physiology have evolved over 54 million years to accommodate a gradual intake ('trickle feeding') of a high-fibre, low-starch, low-fat diet, composed primarily of grasses, rushes, sedges and occasional cereals, over a prolonged feeding period perhaps occupying 16 hours of the day (Houpt, 1990). The stomach would invariably contain significant quantities of ingested fibre with the more dorsal squamous mucosa largely protected from significant exposure to gastric juice. Dietary changes would, most probably, occur naturally in a gradual fashion with changes in the seasons and weather influencing quality and quantity of ingested herbage. Although domestication of the horse probably began approximately 7000 years ago, this is a brief period of time in evolutionary terms. Under modern management systems many horses are fed relatively energy-dense, high-carbohydrate, low-fibre and occasionally high-fat feeds in meals that are often consumed in a relatively short period of time leaving the stomach relatively empty for significant periods of time (Clarke *et al.*, 1990; Houpt, 1990). Furthermore, many differing choices and sources of preserved forage and concentrated feed are available to the horse owner creating a greater likelihood of abrupt dietary changes. It is perhaps no surprise therefore that many epidemiologic studies have found deviations from a 'natural feeding regimen' to be associated with colic (Table 1). Given the presumption that the horse's digestive tract has evolved to suit nutritional conditions experienced by feral horses rather than intensively managed competition horses, the ideal nutritional strategy for avoidance of gastrointestinal disturbances is likely to mimic many aspects of the behaviour and dietary supply of the feral animal (Houpt, 1990). In practice it will frequently prove difficult (or impossible) to exactly mimic a natural diet and dietary behaviour owing to the increased nutritional requirements of working horses and the relatively high nutrient density of available feeds, forages and

Table 1. Examples of odds ratios (OR) and probability values (P) for relative risk of colic derived from mulitivariable analyses in different studies versus referent population.

	OR	P	Reference
Concentrate feeding			
Fed >5 kg/day concentrate	6.3	0.004	Tinker et al., 1997b
Fed >2.7kg/day oats	5.9	0.009	Hudson et al., 2001
Fed 2.5-5 kg/day concentrate	4.8	0.01	Tinker et al., 1997b
Single change in concentrate fed	3.6	<0.001	Tinker et al., 1997b
Change in concentrate within previous 2 weeks	2.6	0.064	Hudson et al., 2001
No whole grain fed	2.5	0.01	Tinker et al., 1997b
Mulitple changes in concentrate fed	2.2	0.02	Tinker et al., 1997b
Concentrate fed (per kg)[1]	1.5	<0.001	Cohen et al., 2006
Grazing			
Change of grazing within previous 2 weeks[3]	29.7	<0.001	Wood et al., 1998
<50% pasture turnout[2]	4.5	<0.01	Cohen et al., 2000
No pasture time[1]	4.0	<0.001	Cohen et al., 2006
No pasture time (or recent decrease)	3.0	0.007	Hudson et al., 2001
Forage			
Hay change within previous 2 weeks	9.8	0.035	Cohen et al., 1999
Change in batch of hay within previous 2 weeks	4.9	<0.001	Hudson et al., 2001
Coastal Bermuda hay fed[4]	4.4	<0.05	Little and Blikslager, 2002
Fed alfalfa hay[2]	4.2	0.01	Cohen et al., 2000
Hay fed from round bales	2.5	0.028	Hudson et al., 2001
Multiple changes in hay fed	2.1	0.01	Tinker et al., 1997b
Miscellaneous			
Diet change within previous 2 weeks	5.0	<0.001	Cohen et al., 1999
Diet change within previous 2 weeks	2.2	<0.001	Cohen et al., 1995
Diet change within previous 2 weeks	2.1	0.005	Cohen and Peloso, 1996

[1] Duodenitis proximal jejunitis cases only.
[2] Enterolithiasis cases only.
[3] Grass sickness cases only.
[4] Ileal impaction cases only.

pasture in comparison to the typical feral diet. Nevertheless, potential dietary risk factors can still be identified and moderated in colic-prone horses.

References

Archer, D.C., Pinchbeck, G.L., Proudman, C.J. and Clough, H.E., 2006. Is equine colic seasonal? Novel application of a model based approach. BMC Vet Res. 24, 27-37.

Archer, D.C. and Proudman C.J., 2006. Epidemiological clues to preventing colic. Vet J. 172, 29–39.

Bell, R.J., Mogg, T.D. and Kingston, J.K. (2007) Equine gastric ulcer syndrome in adult horses: A review. N Z Vet J. 55, 1-12

Cargile, J.L., Burrow, J.A., Kim, I., Cohen, N.D. and Merritt, A.M., 2004. Effect of dietary corn oil supplementation on equine gastric fluid acid, sodium, and prostaglandin E2 content before and during pentagastrin infusion. J Vet Intern Med. 18, 545-549.

Carmalt, J.L., Townsend, H.G.G., Janzen, E.D. and Cymbaluk, N.F., 2004. Effect of dental floating on weight gain, body condition score, feed digestibility, and fecal particle size in pregnant mares. J Am Vet Med Assoc. 225, 1889-1893.

Clarke, L.L., Ganjam, V.K., Fichtenbaum, B., Hatfield, D. and Garner, H.E., 1988. Effect of feeding on renin-angiotensin-aldosterone system of the horse. Am J Physiol. 254, R524-R530.

Clarke, L.L., Roberts, M.C. and Argenzio, R.A., 1990. Feeding and digestive problems in horses. Vet Clin North Am Equine Pract. 6, 319-337.

Cohen, N.D., Gibbs, P.G. and Woods, A.M., 1999. Dietary and other management factors associated with colic in horses. J Am Vet Med Assoc. 215, 53–60.

Cohen, N.D., Matejka, P.L., Honnas, C.M. and Hooper, R.N., 1995. Case-control study of the association between various management factors and development of colic in horses. Texas Equine Colic Study Group. J Am Vet Med Assoc. 206, 667–673.

Cohen, N.D. and Peloso, J.G., 1996. Risk factors for history of previous colic and for chronic, intermittent colic in a population of horses. J Am Vet Med Assoc. 208, 697–703.

Cohen, N.D., Toby, E., Roussel, A.J., Murphey, E.L. and Wang, N., 2006. Are feeding practices associated with duodenitis-proximal jejunitis? Equine Vet J. 38, 526-531.

Cohen, N.D., Vontur, C.A. and Rakestraw, P.C., 2000. Risk factors for enterolithiasis among horses in Texas. J Am Vet Med Assoc. 216, 1787–1794.

Cooper, J.J., Mcall, N., Johnson, S. and Davidson, H.P.B., 2005. The short-term effects of increasing meal frequency on stereotypic behaviour of stabled horses. Appl Anim Behav Sci. 90, 351-364.

Cuddeford, D., 1999. Why feed fibre to the performance horse today? In: Proceedings BEVA specialist days on behaviour and nutrition. Eds P.A.Harris, G.M.Gomarshall, H.P.B.Davidson, R.E.Green. Equine Veterinary Journal Ltd. pp 50-54.

Cuddeford, D., 2000. The significance of pH in the hind gut. Proceedings Dodson and Horrell International Conference on Feeding horses 3, 3-9.

Dabareiner, R.M. and White, N.A., 1995. Large colon impaction in horses: 147 cases (1985-1991). J Am Vet Med Assoc. 206, 679-685.

De Fombelle, A., Julliand, V., Drogoul, C. and Jacotot, E., 2001. Feeding and microbial disorders in horses: 1-Effects of an abrupt incorporation of two levels of barley in a hay diet on microbial profile and activities. J Eq Vet Sci. 26, 439-445.

Drogoul, C., de Fombelle, A. and Julliand, V., 2001. Feeding and microbial disorders in horses: 2: Effect of three hay:grain ratios on digesta passage rate and digestibility in ponies. J Eq Vet Sci. 26, 487-491.

Frank N., Andrews, F.M., Elliott, S.B. and Lew, J., 2005. Effects of dietary oils on the development of gastric ulcers in mares. Am J Vet Res. 66, 2006-2011.

Geor, R.J., 2005. Diet, feeding and gastrointestinal health in horses. Proceedings BEVA & Waltham Nutrition Symposia 1, 89-94.

Getty, S.M., Ellis, D.J., Krehbiel, J.D. and Whitenack, D.L., 1976. Rubberised fencing as a gastrointestinal obstruction in a young horse. Vet Med Small Anim Clin. 71, 221-223.

Goodson, J., Tyznik, W.J., Cline, J.H. and Dehority, B.A., 1988. Effects of an abrupt diet change from hay to concentrate on microbial numbers and physical environment in the cecum of the pony. Appl. Environ Microbiol. 54, 1946-1950.

Green, P. and Tong, J.M.J., 1988. Small intestinal obstruction associated with wood chewing in two horses. Vet Rec. 123, 196-198.

Harris, P.A., 2007. How should we feed horses - and how many times a day? Vet J. 173, 9-10.

Harris, P.A. and Arkell, K., 2005. How understanding the digestive process can help minimize digestive disturbances due to diet and feeding practices. Proceedings BEVA & Waltham Nutrition Symposia 1, 9-14.

Hassel, D.M., Rakestraw, P.C., Gardner, I.A., Spier, S.J. and Snyder, J.R., 2004. Dietary risk factors and colonic pH and mineral concentrations in horses with enterolithiasis. J Vet Intern Med. 18, 346-349.

Hillyer, M.H., Taylor, F.G. and French, N.P., 2001. A cross-sectional study of colic in horses on thoroughbred training premises in the British Isles in 1997. Equine Vet J. 33, 380-385.

Hillyer, M.H., Taylor, F.G., Proudman, C.J., Edwards, G.B., Smith, J.E. and French, N.P., 2002. Case control study to identify risk factors for simple colonic obstruction and distension colic in horses. Equine Vet J. 34, 455-463.

Hintz, H.F., 2000. Equine nutrition update. Proc AAEP 46, 62-79.

Houpt, K.A., 1990. Ingestive behaviour. Vet Clin North Am Equine Pract. 6, 319-337.

Houpt, K.A., Perry, P.J., Hintz, H.F. and Houpt, T.R., 1988. Effect of meal frequency on fluid balance and behavior of ponies. Physiol Behav. 42, 401-407.

Hudson, J.M., Cohen, N.D., Gibbs, P.G. and Thompson, J.A., 2001. Feeding practices associated with colic in horses. J Am Vet Med Assoc. 219, 1419-1425.

Hunter, L.C., Miller, J.K. and Poxton, I.R., 1999. The association of Clostridium botulinum type C with equine grass sickness: a toxicoinfection? Equine Vet J. 31, 492-499.

Hussein, H.S., Vogedes, L.A., Fernandez, G.C. and Frankeny, R.L., 2004. Effects of cereal grain supplementation on apparent digestibility of nutrients and concentrations of fermentation end-products in the feces and serum of horses consuming alfalfa cubes. J Anim Sci. 82, 1986-1996.

Jansen, W.L., Geleen, S.N.J., van der Kuilen, J. and Beynen, A.C., 2002. Dietary soyabean oil depresses the apparent digestibility of fibre in trotters when substituted for an iso-energetic amount of corn starch or glucose. Equine Vet J. 34, 302-305.

Julliand, V., de Fombelle, A., Drogoul, C. and Jacotot, E., 2001. Feeding and microbial disorders in horses: Part 3—Effects of three hay:grain ratios on microbial profile and activities. J Eq Vet Sci. 26, 543-546.

Kaneene, J.B., Miller, R.A., Ross, W.A., Gallagher, K., Marteniuk, J. and Rook, J., 1997. Risk factors for colic in the Michigan (USA) equine population. Prev Vet Med. 30, 23-36.

Kellam, L.L., Johnson, P.J., Kramer, J. and Keegan, K.G., 2000. Gastric impaction and obstruction of the small intestine associated with persimmon phytobezoar in a horse. J Am Vet Med Assoc. 216, 1279–1281.

Little, D. and Blikslager, A.T., 2002. Factors associated with development of ileal impaction in horses with surgical colic: 78 cases (1986–2000). Equine Vet J. 34, 464–468.

Longland, A.C. and Byrd, B.M., 2006. Pasture nonstructural carbohydrates and equine laminitis. J. Nutr. 136, 2099S-2102S.

Longland, A.C. and Cairns, A.J., 2000. Fructans and their implications in the aetiology of laminitis. Proceedings Dodson and Horrell International Conference on Feeding horses 3, 52-55.

Lopes, M.A., White, N.A., Crisman, M.V. and Ward, D.L., 2004. Effects of feeding large amounts of grain on colonic contents and feces in horses. Am J Vet Res. 65, 687-694.

Mackie, R.I. and Wilkins, C.A., 1988. Enumeration of anaerobic bacterial microflora of the equine gastrointestinal tract. Appl Environ Microbiol. 54, 2155-2160.

Mair, T.S., 2002. Small intestinal obstruction caused by a mass of feedblock containing molasses in 4 horses. Equine Vet J. 34, 532-536.

McCarthy, H.E., Proudman, C.J. and French, N.P., 2001. Epidemiology of equine grass sickness: a literature review (1909-1999). Vet Rec. 149, 293-300.

Medina, B., Girard, I. D., Jacotot, E. and Julliand, V., 2002. Effect of a preparation of Saccharomyces cerevisiae on microbial profiles and fermentation patterns in the large intestine of horses fed a high fiber or a high starch diet. J Anim Sci. 80, 2600-2609.

Mehdi, S. and Mohammed, V., 2006. A farm-based prospective study of equine colic incidence and associated risk factors. J Eq Vet Sci. 26, 171-174.

Metayer, N., Lhote, M., Bahr, A., Cohen, N.D., Kim, I., Roussel, A.J. and Julliand, V., 2004. Meal size and starch content affect gastric emptying in horses. Equine Vet J. 36, 436-440.

Morris, D.D., Moore, J.N. and Ward, S., 1989. Comparison of age, sex, breed, history and management in 229 horses with colic. Equine Vet J. (suppl.) 7, 129-132.

Nadeau, J.A., Andrews, F.M., Mathew, A.G., Argenzio, R.A., Blackford, J.T., Sohtell, M. and Saxton, A.M., 2000. Evaluation of diet as a cause of gastric ulcers in horses. Am J Vet Res. 61, 784-790.

Nadeau, J.A., Andrews, F.M., Patton, C.S., Argenzio, R.A., Mathew, A.G. and Saxton, A.M., 2003. Effects of hydrochloric, acetic, butyric, and propionic acids on pathogenesis of ulcers in the nonglandular portion of the stomach of horses. Am J Vet Res. 64, 404-412.

Potter, G.D., Arnold, F.F., Householder, D.D., Hansen, D.H. and Brown, K.M., 1992. Digestion of starch in the small or large intestine of the equine. Europische Konferenz uber die Ernahrung des Pferdes. PferdeHeilkunde Sonderheft 1, 107-111.

Proudman, C.J., 1991. A two year, prospective survey of equine colic in general practice. Equine Vet J. 24, 90–93.

Ragle, C.A., Meagher, D.M., Lacroix, C.A. and Honnas, C.M., 1989. Surgical treatment of sand colic. Results in 40 horses. Vet Surg. 18, 48-51.

Ralston, S.L., Foster, D.L., Divers, T. and Hintz, H.F., 2001. Effect of dental correction on feed digestibility in horses. Equine Vet J. 33, 390-393.

Reeves, M.J., Salman, M.D. and Smith, G., 1996. Risk factors for equine acute abdominal disease

(colic): Results from a multi-center case-control study. Prev Vet Med 26, 285-301.

Tinker, M.K., White, N.A., Lessard, P., Thatcher, C.D., Pelzer, K.D., Davis, B. and Carmel, D.K., 1997a. A prospective study of equine colic incidence and mortality. Equine Vet J. 29, 448–453.

Tinker, M.K., White, N.A., Lessard, P., Thatcher, C.D., Pelzer, K.D., Davis, B. and Carmel, D.K., 1997b. A prospective study of equine colic risk factors. Equine Vet J. 29, 454–458.

Traub-Dargatz, J.L., Kopral, C.A., Seitzinger, A.H., Garber, L.P., Forde, K. and White, N.A., 2001. Estimate of the national incidence of and operation-level risk factors for colic among horses in the United State, spring 1998 to spring 1999. J Am Vet Med Assoc. 219, 67–71.

Traub-Dargatz, J.L., Salman, M.D. and Jones, R.L., 1990. Epidemiologic study of salmonellae shedding in the feces of horses and potential risk factors for development of the infection in hospitalized horses. J Am Vet Med Assoc. 196, 1617–1622.

Uhlinger, C., 1992. Investigations into the incidence of field colic. Equine Vet J. (suppl.) 13, 16-18.

van Weyenberg, S., Buyse, J. and Janssens, G.P., 2007. Digestibility of a complete ration in horses fed once or three times a day and correlation with key blood parameters. Vet J. 173, 311-316.

Weese, J.S., 2002. Microbiologic evaluation of commercial probiotics. J Am Vet Med Assoc. 220, 794-797.

Weese, J.S., Anderson, M.E., Lowe, A., Penno, R., da Costa, T.M., Button, L. and Goth, K.C., 2004. Screening of the equine intestinal microflora for potential probiotic organisms. Equine Vet J. 36, 351-355.

Weese, J.S. and Rousseau, J., 2005. Evaluation of Lactobacillus pentosus WE7 for prevention of diarrhea in neonatal foals. J Am Vet Med Assoc. 226, 2031-2034.

Wood, J.L.N., Milne, E.M. and Doxey, D.L., 1998. A case-control study of grass sickness (equine dysautonomia) in the United Kingdom. Vet J. 156, 7-14.

Recent papers on equine nutrition related to medicine (2005-06)

Simon R. Bailey

Faculty of Veterinary Science, University of Melbourne, Victoria 3010, Australia

The following five papers illustrate some of the latest research in a number of important areas where nutrition may have a significant impact on equine medicine. These papers include the subjects of nutritional risk factors for laminitis and the treatment of obesity in ponies, the role of nutritional factors in the pathogenesis of intestinal disease and neuronal disease, and parenteral nutrition in foals. They illustrate some basic principles of nutrition as well as the importance of research in developing evidence-based medicine and how nutrition may play a key role in the pathogenesis of several common clinical diseases.

Treiber, K.H., Kronfeld, D.S., Hess, T.M., Byrd, B.M., Splan, R.K. and Staniar, W.B., 2006. Evaluation of genetic and metabolic predispositions and nutritional risk factors for pasture-associated laminitis in ponies. J Am Vet Med Assoc 228:1538-1545.

Objective of the study

To evaluate genetic and metabolic predispositions and nutritional risk factors for development of pasture-associated laminitis in ponies.

Introduction

While pony breeds are likely to suffer more commonly from pasture-associated acute laminitis than horses, it is being increasingly recognised that certain individuals within a herd are predisposed to the condition. Identifying individual animals at risk, before they suffer serious episodes of laminitis, will potentially be of great benefit;

bearing in mind the difficulties in treating and managing recurrent and chronic laminitis. Insulin resistance seems to share a strong association with laminitis risk, but this has not previously been investigated in a large population of ponies. The highest incidence of laminitis occurs in spring/summer, when starch and fructan carbohydrate concentrations in the pasture are greatest – this may exacerbate insulin resistance.

Methods

An inbred herd of 160 Welsh and Dartmoor ponies were studied, in which 54 ponies had previously been diagnosed with laminitis. A genetic link for laminitis incidence was investigated in these ponies using pedigree analysis. Body condition was assessed, and blood samples taken to determine insulin sensitivity.

Main findings

There appeared to be a strong genetic link between those ponies developing laminitis, with a dominant mode of inheritance. Laminitis-predisposed ponies exhibited increased body condition scores, plasma triglycerides, insulin secretory responses and decreased insulin sensitivity. These criteria could be used to provide a predictive index which could predict laminitis risk with a power of 78%.

Practical interest

Prelaminitic metabolic syndrome in apparently healthy ponies seems to be comparable to metabolic syndrome in humans. Insulin resistance and dyslipidaemia are two features of the condition and unpublished data (Bailey *et al.*) suggests that syndrome may also include mild hypertension. Considering several of these factors together may enable some prediction of laminitis risk associated with this syndrome. Prelaminitic metabolic syndrome identifies ponies requiring special management, such as avoiding high starch intake that exacerbates insulin resistance. Also, in this inbred pony herd there appears to be a clear genetic component to this syndrome, which raises hope of identifying the genes involved and the development of genetic screening tests.

Comments

It should be noted that the situation in the wider pony population may not be as clear-cut as in this closed herd. It is likely that any genetic predisposition to laminitis in outbred ponies involves several genes and complicated phenotypic differences. This study also highlights the fact that using single criteria, particularly basal insulin and glucose concentrations, may show statistically significant differences when comparing large groups of normal and laminitis-predisposed ponies, but is less useful for diagnosing this syndrome in individual animals. This is because there is considerable overlap between the two groups. Therefore, consideration of multiple criteria gives more predictive power. Dynamic tests of the insulin secretory response may eventually give further sensitivity.

Krause, J.B. and McKenzie, H.C., 2007. Parenteral nutrition in foals: a retrospective study of 45 cases (2000-2004). Equine Vet J 39(1):74-8.

Objective of the study

The administration of parenteral nutrition (PN) to critically ill foals was examined retrospectively to determine the effects of PN formulation and variables on the incidence of PN-associated complications and outcome.

Introduction

Adequate nutritional support of sick foals in critical care is an important aspect of treatment. When enteral feeding is contraindicated, parenteral nutrition (PN) provides a source of energy and protein. If sufficient nutritional support is not provided, affected foals quickly develop a negative energy balance which results in protein catabolism in order to utilise amino acids for energy production. In order to meet the maintenance requirements of sick neonatal foals, a gross energy intake of 260-289 kJ/kg bwt/day is recommended, with 544-607 kJ/kg bwt/day required for growth. Formulating PN with lipids enables the provision of a larger number of Joules per unit volume compared with dextrose. However, this could increase the risk of hyperlipidaemia and bacterial contamination of the formulation. This is the first study to critically assess the use of PN in a large group of foals.

Methods

Medical records of 45 foals presented at a referral centre which received PN were reviewed for the years 2000-2004. This was to establish the effect of PN formula on the occurrence and type of complications and clinical outcome, and also the effect of disease severity on these variables. Sick neonatal foals were scored by sepsis score, blood cell count and biochemistry. Two basic PN formulae were used, either composed of 50% dextrose and 8.5% amino acid solutions, or 50% dextrose, 8.5% amino acid and 20% lipid solutions. Mean kJ/kg bwt/day was 173 for the lipid-free PN solution and 263 for the lipid containing solution.

Main findings

Formulation of PN was not associated with the development of complications, and there was no association of PN formula with patient survival. Disease severity was positively associated with the development of PN complications and the occurrence of PN complications was associated with a decreased likelihood of survival.

Practical interest

The use of lipid-containing PN solutions facilitates the delivery of energy to the critically ill foal without increasing the risk of deleterious side effects. Severely ill foals are more prone to develop complications associated with PN and to have a poor outcome.

Comment

Influences of PN on weight gain and body condition were not included in this study because of incomplete data. The recording of weight gain is an important aspect of monitoring the adequate supply of nutrients. Even if foals receiving PN might not gain weight, it is important to monitor whether they can at least maintain their weight or if they are in negative energy balance.

The foals were selected to receive PN based on severity of their clinical condition. Therefore there was no age and severity matched control group available for comparison. To evaluate the effect of PN administration on outcome, a prospective randomised study would be required to compare foals with the same type of disease treated and not treated with PN.

Cohen, N.D., Toby, E., Roussel, A.J., Murphey, E.L. and Wang, N., 2006. Are feeding practices associated with duodenitis-proximal jejunitis? Equine Vet J 38(6):526-31.

Objective of the study

A case controlled study to determine whether there is evidence that feeding practices are associated with increased risk of duodenitis-proximal jejunitis (DPJ).

Introduction

DPJ is an idiopathic syndrome of small intestinal ileus characterised by copious nasogastric reflux associated with gross and microscopic pathological changes restricted typically to the duodenum, proximal jejunum or both. The cause remains unknown, but toxic and infectious agents have been implicated. *Salmonella spp.* Have been isolated from gastric fluid and faeces of affected horses, as has *Clostridium perfringens.* However, these organisms are not isolated consistently from all cases of DPJ, and can sometimes be isolated from healthy horses. Aflatoxicosis may also produce haemorrhagic enteritic lesions similar to DPJ. Endotoxin has been isolated from the blood of horses with DPJ but experimentally induced toxicosis does not consistently result in a syndrome identical to DPJ. Feeding a diet with a large proportion of the energy content derived from concentrate has been putatively associated with DPJ, but this has not been previously evaluated using a controlled study design.

Simon R. Bailey

Methods

Feeding practices of cases of DPJ diagnosed at Texas A&M University between 1997 and 2003 were compared with those of 2 populations of control horses (colic controls and lameness controls) admitted to the clinic from the same period. Comparisons were made using logistic regression. Cases were defined as horses with a clinical diagnosis of DPJ and produced excessive nasogastric reflux (>20 l/day) for >24h from time of onset of clinical signs. Horses were excluded if they were found to have a mechanical obstruction of the intestinal tract, or if they recovered within 24h. No ponies or animals <1 year old were included. Type and amount of concentrate and roughage in each horse's normal diet was recorded, and whether horses received a proportion of their roughage from grazing.

Main findings

70 cases of DPJ were identified which met the criteria and had dietary data, together with 153 colic controls and 108 lameness controls. There were no significant differences between DPJ cases and either control group regarding whether a texturised (sweet-feed) concentrate, pelleted concentrate, oats or other concentrate, coastal Bermuda hay or alfalfa hay was fed routinely. There was a significant difference between groups in the amount of total concentrate fed, with DPJ horses receiving 4.1 kg (median; IQ range 1.8-4.8) compared with 2.7 kg in both control groups. Horses with DPJ were also more likely to have had history of grazing pasture than horses in either control group.

Practical interest

Horses with DPJ were fed significantly more concentrate and were significantly more likely to have grazed pasture than either control populations. Therefore feeding and grazing practices differ among horses with DPJ relative to horses with other forms of colic, or horses with lameness. These results substantiate clinical impressions and results described in review articles and case series. However, the association of DPJ with pasture grazing as a portion of the diet has not been previously described. This may provide some further clues to the aetiology of this condition.

Applied equine nutrition and training

Comment

The associations were not sufficiently strong to merit diagnostic or predictive application. Furthermore, the results of the study should be interpreted with caution for several reasons. Firstly, these data were collected retrospectively and no attempt was made to validate owner-reported histories recorded in medical charts. Secondly, no information was available about temporal changes in feeding and whether there had been any recent changes in diet that might have further explained or confounded the results. Also, to minimise misclassification of cases, milder cases of DPJ were excluded from the analysis.

Buff, P.R., Johnson, P.J., Wiedmeyer, C.E., Ganjam, V.K., Messer, N,T. and Keisler, D.H., 2006. Modulation of leptin, insulin, and growth hormone in obese pony mares under chronic nutritional restriction and supplementation with ractopamine hydrochloride. Veterinary Therapeutics 2006 7(1):64-72.

Objective of the study

To determine the effects of a negative energy balance, achieved by feeding obese pony mares at 75% of ad libitum feed intake, and the effects of treating with ractopamine hydrochloride, over a period of 6 weeks.

Introduction

Horses that are fed beyond their nutritional requirement and are physically inactive will develop obesity, which is often accompanied by insulin resistance and an increased risk of laminitis. It has been postulated that some ponies are more metabolically efficient ('easy keepers') than others, having retained the ability to cope when food is scarce by developing additional body fat. Therefore under modern management conditions, the combination of feeding starch-rich rations and protracted periods of stall confinement can lead to the acquisition and maintenance of substantial adiposity in these animals. Recently, seasonal variations of hormones associated with appetite,

Simon R. Bailey

including the adipokine, leptin, have been reported in horses, which may support this theory.

Ractopamine hydrochloride is a β-adrenergic agonist which has recently been approved for use in finishing pig rations in the US, for augmenting weight gain and carcass leanness. It was hypothesised that this compound may have some advantages in reducing adiposity over the β$_2$ adrenergic agonist, clenbuterol, which is already known to have a partitioning effect in horses.

Methods

Fifteen obese pony mares (condition scores greater than 7 out of 9) were used, divided into 3 treatment groups of equal body weight range (mean 250 kg). One group acted as control, the other groups received either 0.6 or 1.0 mg/kg/day of ractopamine (divided into two daily doses) throughout the 6 weeks of the study. The mares were offered ad libitum brome hay for 5 days to determine ad libitum intake, and subsequently each mare was fed at 75% of this intake by weight, but including 10% by weight of oats coated with molasses to act as a carrier for the ractopamine. Blood samples were collected periodically during the 6 week study for measurement of leptin, insulin and growth hormone.

Main findings

This study showed that reducing feed intake of brome grass hay to 75% of ad libitum intake in obese pony mares reduced bodyweight without induced exercise. There was a steady decline in body weights observed in all mares, regardless of treatment. Supplementation of ractopamine hydrochloride at 1.0 mg/kg/day resulted in a tendency for increased weight loss in the first two weeks, however this did not reach statistical significance (P = 0.09). Thereafter the body weights decreased at similar rates in all 3 groups. Modulation of the obesity-associated hormones leptin and insulin was observed as a result of caloric restriction. Plasma insulin declined markedly in all 3 groups during the first 3 weeks of the study and remained low. Growth hormone decreased during the first week and then recovered to baseline levels, while leptin showed small decreases after 4 days and two weeks, with recoveries in between.

Applied equine nutrition and training

Practical interest

The use of pharmacological agents in combination with nutritional restriction may promote weight loss in obese horses, when perhaps they are unable to exercise because of laminitic pain. However, this study would suggest that the effects of caloric restriction are not enhanced by the use of ractopamine. Weight loss will also decrease plasma insulin (obesity is associated with insulin resistance and laminitis risk). It should be stressed that care should be taken to avoid the risk of hyperlipaemia when restricting food intake in obese ponies.

Comments

In the present study, ractopamine did not have any significant effects on the rate of weight loss, although clenbuterol has been previously shown to reduce adiposity in horses and act as a repartitioning agent. Ractopamine is primarily a β_1-adrenoceptor agonist, whereas clenbuterol activates β_2 adrenoceptors. Ractopamine seems to have a high affinity for subcutaneous adipose tissue in pigs, but no such studies have been carried out in horses. Unfortunately, adiposity was not measured in the present study, and weight loss may not have been the best measurement of metabolic changes, if lean muscle mass was increasing at the same time as adipose tissue was decreasing. Therefore, further studies would need to be carried out with this compound to establish efficacy. Furthermore, a direct comparison with clenbuterol would be useful in a future study.

L-thyroxine has been found to reduce plasma lipid concentrations and improve insulin sensitivity in healthy horses (Frank *et al.*, 2005). Therefore, this may prove to be a more useful alternative to β agonists for promoting weight loss in obese ponies and horses.

References

Frank,N., Sommardahl, C.S., Eiler, H., Webb, L.L., Denhart, J.W. and Boston, R.C., 2005. Effects of oral administration of levothyroxine sodium on concentrations of plasma lipids, concentration and composition of very-low-density lipoproteins, and glucose dynamics in healthy adult mares. Am J Vet Res 66:1032-1038

Simon R. Bailey

Divers, T.J., Cummings, J.E., de Lahunta, A., Hintz, H.F. and
Mohammed, H.O., 2006. Evaluation of the risk of motor neuron
disease in horses fed a diet low in vitamin E and high in copper
and iron. Am J Vet Res 67(1):120-6.

Objective of the study

To determine whether equine motor neuron disease (EMND) could be
induced in adult horses fed a diet low in vitamin E and high in copper
and iron.

Introduction

EMND is a naturally occurring neurodegenerative disease of the somatic
lower motor neuron system in adult horses. Clinical signs include
weight loss from muscle wasting, trembling, muscle fasciculations
and prolonged recumbency. Accumulation of lipopigment in spinal
cord capillaries and the predilection for denervation of the highly
oxidative type I muscle fibres suggest that EMND is an oxidative
disorder. Epidemiological studies indicate that absence of green forage
for at least 18 months, and high grain diets are risk factors for this
condition. Low plasma concentrations of the antioxidant vitamin E
has been a consistent finding in clinical cases. Vitamin E deficiency
is also associated with the diffuse neurodegenerative disorder of
white matter neurons in young growing equids, equine degenerative
myeloencephalopathy (EDM). Vitamin E exists in 8 naturally occurring
forms, with α-tocopherol being the most important and active. Green
forages are the major source of naturally occurring vitamin E, whereas
cereal grains contain minimal amounts.

Increased concentrations of copper have also been recorded in the
spinal cords of affected animals with EMND. Increased hepatic iron
has also been reported in some animals. Both copper and iron can act
as pro-oxidants through the Fenton reaction, to yield free hydroxyl
ions.

Methods

Horses in the experimental group (n = 8) were confined to a dirt lot
and fed a concentrate low in vitamin E (approx 20 U/kg of diet; NRC

recommendation 50 U/kg diet) and high in copper (4,000 ppm; >10 times recommended NRC intake) and iron (2,000 ppm; >5 times the NRC recommended intake) in addition to grass hay that had been stored for 1 year. Control horses (n = 51) were fed the same basic concentrate (mainly cracked corn, whole oats, soybean meal, with molasses, limestone, and dicalcium phosphate) but containing NRC-recommended amounts of copper, iron and vitamin E (30 U of D-1 α-tocopherol acetate/kg). Control horses only also had seasonal access to pasture. Horses that developed clinical signs of EMND were euthanased along with age-matched control horses to determine differences in hepatic concentrations of vitamin E, vitamin A, copper, iron and selenium.

Main findings

Four experimental horses developed clinical signs of EMND after 21-28 months on the diet, with no clinical signs being noted before this time. Clinical signs included acute onset trembling, fasciculations, shifting of weight when standing, excessive recumbency and weight loss. Histologic changes in the spinal cord, peripheral nerves and muscles were characteristic of EMND. Plasma concentrations of vitamin E decreased in all 8 experimental horses. Values at euthanasia or at the end of the study were 0.09-0.51 μg/ml. There were no significant changes in plasma concentrations of vitamin A, selenium, copper or serum concentrations of ferritin. There were no significant differences in these analytes between experimental horses with EMND and those 4 that did not develop EMND. No control horses developed EMND.

Practical interest

These results provide further evidence that lack of access to pasture, dietary deficiency of vitamin E or excessive dietary copper are likely risk factors for EMND. Tissue stores of vitamin E are abundant in horses that have seasonal access to green forage, and several months of feeding a vitamin E deficient diet are required to develop a deficiency severe enough for oxidative injury and signs of EMND. Not all horses at risk developed EMND, suggesting that there could be individual susceptibility to a disturbed antioxidant-pro-oxidant balance (oxidative stress). The role of copper and iron as pro-oxidants is less clear. Copper concentrations were high in the 4 horses with clinical signs of EMND, but high copper concentrations have not been detected in horses with naturally occurring EMND. Horses may be relatively resistant to copper-induced hepatopathy. Increased hepatic concentrations of iron have been noted in horses with naturally occurring EMND, but iron concentrations are not high in the spinal cords of these horses.

Comments

It should be noted that the synthetic vitamin E added to concentrates is usually D-1 α-tocopherol acetate (also known as *all-rac* α-tocopherol acetate), which has a lower bioavailability and lower potency than natural vitamin E (RRR-α-tocopherol). Vitamin E was not measured in the spinal cord tissue of the horses in the present study, but in humans tocopherol concentrations in CSF correlate well with serum concentrations. Previous studies have shown that feeding a vitamin-E deficient diet for 3 months caused no ill effect, therefore it is only prolonged feeding of such a diet that increases the risk of EMND. It is recommended that all horses without access to green forage for prolonged periods should receive dietary supplementation with vitamin E to ensure a daily intake of at least 1 U/kg bodyweight per day, irrespective of age.

Diagnosis and management of insulin resistance and Equine Metabolic Syndrome (EMS) in horses

Nicholas Frank
University of Tennessee, TN
University of Nottingham, UK

Introduction

The term 'equine metabolic syndrome (EMS)' has been adopted to describe a collection of clinical signs that contribute to the development of insidious-onset laminitis in horses. Equine metabolic syndrome has not been rigorously defined and critics justifiably argue that more studies are required to substantiate its existence. However, it is still useful to recognize this clinical syndrome because horses and ponies with EMS are at higher risk for developing laminitis, and effective management of this condition appears to aid in the prevention of laminitis.

Definition

It is likely that the definition of EMS will be revised as we learn more, but this syndrome is currently defined by the presence of (1) insulin resistance (IR), (2) obesity and/or regional adiposity, and (3) prior or current laminitis. Evidence of prior laminitis comes from the history provided by the client or by detection of divergent growth rings on the hooves (founder lines), which are assumed to result from prior subclinical laminitis episodes. Some affected horses show radiographic evidence of third phalanx rotation, but are not lame at the time of examination. One key feature of EMS is the underlying metabolic status of the horse or pony. This concept is easily understood by horse owners because they often describe their horse as an 'easy keeper', but much more difficult to define scientifically.

Metabolic Syndrome has been used to describe this syndrome in the past (Johnson, 2002), but it is preferable to use the descriptor 'equine' when referring to horses and ponies. Metabolic Syndrome is a term used in human medicine to describe a set of risk factors identified in people that are at higher risk for coronary heart disease, stroke, or diabetes. In contrast, EMS describes a clinical syndrome that is unique to the equine species because of its connection with laminitis.

Obesity and laminitis have also been attributed to hypothyroidism in the past, but we now recognize that low resting thyroid hormone concentrations accompany extrathyroidal illness in horses (Breuhaus *et al.*, 2006), and hypothyroidism should ideally be diagnosed by performing hormone challenges. When low resting total triiodothyronine (tT_3) and total thyroxine (tT_4) concentrations are detected in obese insulin resistant horses, they are more likely to be a consequence rather than a cause of systemic problems.

Clinical presentation

Based upon our experience, EMS is most common in pony breeds, Morgans, Paso Finos, and Norwegian Fjords. We have also examined Arabians, Quarter Horses, Saddlebreds, Tennessee Walking Horses, Thoroughbreds, and Warmbloods with this condition, indicating that a number of breed groups are represented (Frank *et al.*, 2006). When viewed simplistically, easy keeper breeds are most commonly affected, whereas hard keeper breeds such as Thoroughbreds and Standardbreds are less likely to develop EMS.

Susceptibility to EMS is likely to be established from birth and some horses develop obesity at 3 or 4 years of age. However, most horses are between 5 and 15 years of age when veterinary or farrier services are first required because of laminitis. It is interesting to speculate that EMS is only recognized at this later age because horses must remain obese for several years before IR develops. Alternatively, the susceptibility to laminitis may increase over time in horses that are chronically insulin resistant.

Many horses are out on pasture when laminitis is first detected, and this occurs more frequently in the spring when the pasture has gone through a period of rapid growth. Upon closer examination, founder lines are sometimes present on the hoof surfaces, which may indicate subclinical laminitis. Other horses with EMS are first recognized when they present with colic, hyperlipemia, or reproductive problems. Horses with EMS appear to be at greater risk for colic caused by pedunculated lipomas, and this problem may develop at an earlier age in affected animals. Obesity has also been associated with abnormal reproductive cycling in mares (Vick *et al.*, 2006).

Physical characteristics of EMS include generalized obesity and/or regional adiposity. Presence of a 'cresty neck' is the most important form of regional adiposity in horses and mean neck circumference was negatively correlated with insulin sensitivity in a small group of obese insulin resistant horses that we studied (Frank *et al.*, 2006). Noticeable fat deposits are sometimes found close to the tailhead, in the sheath, within the supraorbital fossae, and occasionally as randomly distributed subcutaneous masses throughout the trunk region.

Defining insulin resistance

Insulin resistance can be broadly defined as a decrease in tissue responses to circulating insulin, which causes a decrease in insulin-mediated glucose uptake into skeletal muscle, adipose, and liver tissues (Kronfeld *et al.*, 2005). When using the combined glucose-insulin test (CGIT) described below, IR is recognized by the slower decline in blood glucose concentrations following exogenous insulin administration, indicating that insulin-sensitive tissues are less responsive to the hormone (Eiler *et al.*, 2005). Compensated IR is the most common glucose and insulin metabolism abnormality identified in horses and ponies (Treiber *et al.*, 2006a, b). In this situation, serum insulin concentrations are higher than normal because more insulin is secreted from the pancreas to compensate for lower tissue responses. Compensated IR is thought to lead to uncompensated IR if pancreatic insufficiency develops. This is sometimes also referred to as beta cell exhaustion and is recognized when serum insulin concentrations are lower than expected for the blood glucose concentrations detected. Uncompensated IR can lead to type 2 diabetes mellitus if blood glucose levels rise above the renal threshold and glucosuria develops. In horses that we have evaluated, compensated IR is most common abnormality detected, whereas uncompensated IR is rare and usually associated with advanced pituitary pars intermedia dysfunction (PPID; also called equine Cushing's disease) (Treiber *et al.*, 2006a; 2005). Diabetes mellitus is very rare in horses (Johnson *et al.*, 2005).

Insulin resistance is sometimes, but not always associated with PPID in horses. Glucocorticoids inhibit glucose uptake into insulin-sensitive tissues such as skeletal muscle and adipose tissue, and stimulate gluconeogenesis within the liver. This results in hyperglycemia, which preserves glucose delivery to high priority tissues such as the brain during times of stress or danger. It is presumed that PPID induces chronic IR through excessive cortisol production, but studies have not been performed to determine the relative importance of adrenocorticotropin hormone (ACTH) itself and other pituitary products such as corticotrophin-like intermediate peptide (CLIP) in horses with PPID. Uncompensated IR and glucosuria are occasionally detected in patients with advanced PPID.

Physiological versus pathological insulin resistance

Known physiological causes of IR include stress, administration of exogenous corticosteroids, and gestation. Administration of dexamethasone induces IR and exogenous epinephrine inhibits glucose uptake after dextrose administration in horses (Tiley *et al.*, 2006; Geor *et al.*, 2000). Dexamethasone-induced IR may result from alterations in both insulin signal transduction and glucose transporters (Tiley *et al.*, 2006; Patel *et al.*, 2006). Pregnancy enhances the pancreatic response to glucose and lowers insulin sensitivity in mares. Fowden *et al.* (1984) detected exaggerated pancreatic insulin responses to exogenous glucose and feeding when pregnant mares were compared with non-pregnant animals. Insulin sensitivity decreases during pregnancy to ensure adequate delivery of glucose to the placenta and fetus. Glucose delivery to the placenta occurs along a concentration gradient that is not affected by insulin, so reductions in insulin sensitivity lower maternal glucose utilization. This ensures that sufficient glucose is available for the developing fetus, which gains 45% of its final weight after 270 days of gestation (Fowden *et al.*, 1984).

Obesity and insulin resistance

Obesity and IR appear to be associated in horses, but a direct cause-and-effect relationship has not been established. In a recent study of 300 mature (4 to 20 years of age) client-owned horses in Virginia, the prevalence of hyperinsulinemia (> 30 mU/L) was 10%, and 18 of 30 affected horses were judged to be obese on the basis of body condition score (personal communication, Drs. Ray Geor, Scott Pleasant, and Craig Thatcher). Results of this study suggest that obesity and IR are associated, but also demonstrate that IR can occur in leaner horses. Relationships between obesity and IR require further study and dynamic testing may be required to thoroughly investigate this association. Mild IR may have gone undetected in the aforementioned study because only resting glucose and insulin concentrations were measured. It is likely that more sensitive measures will be required if there is an initial decline in insulin sensitivity that accompanies the development of obesity in horses. It should also be considered whether the duration of obesity plays a role in the development of IR in horses. Chronically obese horses and ponies are likely to be at greater risk for developing IR.

Insulin resistance in leaner horses with regional adiposity

Patients fitting this description are more challenging to understand because IR is not accompanied by obesity. However, affected horses still exhibit regional adiposity in the form of a cresty neck, bulging supraorbital fat, and enlarged fat pads close to the tail base, or have subcutaneous adipose tissue masses randomly distributed throughout the trunk region. It is well established in human medicine that the accumulation of fat within the abdomen is a risk factor for IR and coronary heart disease, and this is sometimes referred to as abdominal, omental, or central obesity (Johnson *et al.*, 2002; Utzschneider *et al.*, 2004). Increased 11β-hydroxysteroid dehydrogenase 1 (11β-HSD1) activity within these tissues is thought to raise local cortisol concentrations above normal and alter adipocyte physiology (Johnson *et al.*, 2002). This condition has been referred to as peripheral Cushing's disease in humans and it has been suggested that 11β-HSD1 activity is higher in the adipose tissues of horses with EMS (Johnson *et al.*, 2002). However, this theory has not been substantiated in horses and it should be recognized that increased 11β-HSD1 activity may be a consequence, rather than a cause of IR.

A thinner body condition is more common in older horses with PPID and EMS may predispose horses to PPID. Horses that are being well managed through diet and exercise will also look thinner, but retain the regional adiposity.

Pathophysiology of equine metabolic syndrome

Equine metabolic syndrome likely develops as genetic and environmental factors interact. In a recent study, Treiber *et al.* (2006a) detected IR in ponies and identified a heritable pattern to this condition within the herd studied. We have also examined horses with EMS that were directly related. However, it is sometimes difficult to determine whether these animals share a common genetic defect or are simply being exposed to the same risk factors, such as the overfeeding of concentrates. The easy keeper concept is relevant to this issue of genetic susceptibility. Certain breeds or genetic lines may have undergone evolutionary adaptations to survive in harsher environments, and these horses or ponies may be more efficient at converting poorer quality forages into energy. Environmental factors that contribute to the development of EMS therefore include grazing on lush pasture, concentrate feeding, and interference with seasonal weight loss.

Feeding excessive calories to genetically susceptible horses likely predisposes them to obesity and for some EMS patients there is a clear history of overfeeding. Obesity sometimes results from owners feeding concentrates to horses that are already at acceptable body weight after grazing on pasture. Some concentrates such as sweet feed, induce weight gain and potentially exacerbate IR. These feeds are usually rich in starches and simple sugars (e.g. molasses) and create a glycemic response that may be more difficult for the insulin resistant horse to cope with. It should also be recognized that the desired body condition for horses varies markedly from one horse owner to another and may differ depending upon the breed. It must also be recognized that good quality pastures can sometimes provide too many calories for the genetically susceptible horse. These horses and ponies are evolutionarily adapted to grazing on poorer forages and readily gain weight when out on pasture.

Finally, it must be questioned whether human interventions interfere with natural protective mechanisms in the horse. Range horses would be expected to lose weight during the winter months when forages are scarce and poorer in quality. However, it is common practice to feed more concentrates at colder times of year to offset this weight loss, even though this intervention may prevent natural correction of IR

Relationships between obesity and IR are complex. *Not all obese horses are insulin resistant and IR is not always accompanied by obesity.* Some horses may be more tolerant of obesity or this condition may have to persist for extended periods of time before IR develops. It is thought that obesity and IR are linked through the disruption of cellular functions by intracellular lipid. Insulin resistance develops as lipid accumulates within myocytes and adipocytes and interferes with insulin receptors, insulin signaling pathways, or glucose transporters. Adipose tissues also expand as the number and size of adipocytes increases, and these tissues release greater more adipokines including leptin, adiponectin, and resistin that act locally and enter the circulation. Adipokines and inflammatory cytokines such as interleukin-6 and tumor necrosis factor-α that are released from macrophages within adipose tissues can exert pro-inflammatory effects that may contribute to the development of IR.

Relationship with pituitary pars intermedia dysfunction

Relationships between the hypothalamic-pituitary gland-adrenal gland axis and EMS must still be elucidated. Some clinicians propose that EMS is simply an early manifestation of PPID, but it is more likely that chronic obesity and IR increase the risk of pituitary dysfunction in horses. Pituitary pars intermedia dysfunction seems to develop at a younger age in horses with EMS and these animals go through a *transition state*. Easy keeper horses that were once obese start to look thinner and require more calories for maintenance. Skeletal muscle atrophy and hirsutism may not be evident at this time, but shedding can be delayed and there is a detectable shift in metabolism. Examination of these relationships is hampered by the limitations of currently available diagnostic tests for PPID. These tests, including the dexamethasone suppression test and resting plasma adrenocorticotropin hormone (ACTH) concentrations, do not detect mild or early disease. It is therefore necessary to rely more upon clinical judgment in these cases, and consider pergolide therapy without confirmatory test results.

Insulin resistance and laminitis susceptibility

All of the pieces of the puzzle must be assembled before we can fully understand the association between IR and laminitis in horses and ponies. However, there are three broad mechanisms by which IR could predispose horses to laminitis: (1) impaired glucose delivery to hoof keratinocytes, (2) altered blood flow or endothelial cell function within the vessels of the foot, or (3) development of a pro-inflammatory or pro-oxidative state. The first theory is supported by results of a study performed by Pass *et al.* (1998) in which it was demonstrated that hoof tissue explants separated at the dermal-epidermal junction when deprived of glucose. Furthermore, Mobasheri *et al.* (2004) determined that GLUT4 proteins are found in equine keratinocytes, which suggests that insulin-stimulated glucose uptake occurs in the hoof. Studies examining the relationship between IR and blood flow have not been performed to date in horses, but insulin is known to act as a slow vasodilator in humans, and IR has been associated with a decrease in peripheral vasodilation and endothelial cell dysfunction (Rask-Madsen *et al.*, 2007; Yki-Jarvinen and Westerbacka, 2000).

If IR is a determinant of susceptibility to pasture-associated laminitis, then what triggers the laminitis episode itself? It appears that the nonstructural carbohydrate (NSC) content of the pasture grass plays an important role in this process. Nonstructural carbohydrates include simple sugars, starch, and fructans (polymers of fructose), and levels of these components vary considerably within grass according to geographical location, soil type, weather conditions, and time of day (Hoffman *et al.*, 2001). These carbohydrates are likely to affect the susceptible horse in two ways. First, excessive sugar consumption could exacerbate IR as it does in diabetic humans and second, consumption of large quantities of NSC might trigger alterations in bacterial flora within the large intestine. Such alterations in large intestinal bacterial flora have been induced by orally administering oligofructose (a fructan) to horses (Al Jassim *et al.*, 2005). Increased fermentation and changes in bacterial flora lower the intraluminal pH, which increases intestinal permeability. This altered intestinal environment may also lead to increased production of triggering factors for laminitis. Potential triggering factors include exotoxins, endotoxins, and vasoactive amines, and the increase in intestinal permeability allows these factors to pass more easily into the circulation (Bailey *et al.*, 2002).

Testing procedures

Resting serum insulin concentrations

This is the easiest measurement to perform and is a useful screening test because compensatory hyperinsulinemia is a common feature of IR in horses. Pancreatic insulin secretion increases to compensate for the decrease in tissue effectiveness, so resting serum insulin concentrations are elevated in horses with moderate or severe IR. However, there are two situations where this test will not be helpful. In the first situation, hyperinsulinemia has not developed yet because only mild IR is present, and in the second situation, pancreatic insufficiency has developed as a consequence of prolonged disease.

Blood samples must be collected from horses after they have been held off pasture for at least 12 hours and fed hay overnight. Grazing on pasture can raise serum insulin concentrations if the sugar concentrations are high in the grass, and grain will cause a peak in insulin levels for a few hours after the meal has been consumed. Detection of IR on the basis of resting insulin concentrations is hindered by the wide reference ranges provided by laboratories. The upper limit of the insulin reference range is 30 µU/mL (mU/L) at the University of Tennessee, but we consider > 20 µU/mL to be suggestive of IR. High-normal glucose concentrations (> 100 mg/dL or 5.5 mmol/L; multiply by 18 to convert units) are also detected in some horses with IR.

Pain and stress associated with acute laminitis markedly elevate resting serum insulin concentrations in EMS patients. Resting serum insulin concentrations can often range from 100 to 400 µU/mL in horses and ponies with clinical laminitis. It is therefore necessary to reevaluate these patients several weeks later after the pain of laminitis has subsided.

Nicholas Frank

Combined glucose-insulin test (CGIT)

This dynamic test provides a better estimation of insulin sensitivity and can detect IR in patients with resting serum insulin concentrations that are within reference ranges. The time taken for blood glucose concentrations to return to baseline is recorded as a measure of insulin sensitivity and serves as a reference point to compare to when testing is repeated after management practices have been instituted. Horses should be held off pasture and fed only hay the night before the test. Hay can also be fed free choice during the test and this will help to keep horses calm during the procedure. The intravenous catheter should ideally be placed the night before testing to minimize the confounding effects of stress, but quieter horses can be tested on the same day.

A pre-infusion (baseline) blood sample is collected and then 150 mg/kg body weight (bwt) 50% dextrose solution is infused, immediately followed by 0.10 units/kg bwt regular insulin (Humulin R®; Eli Lilly, Indianapolis, IN) (Eiler *et al.*, 2005). These dosages are equivalent to 150 mL of 50% dextrose (500 mg/mL) and 0.50 mL of regular insulin (100 units/mL) for a horse weighing 500 kg. Insulin should be drawn into a tuberculin syringe and then transferred into a larger syringe containing 1.5-mL sterile saline (0.9% NaCl) prior to infusion. Blood samples are collected at 1, 5, 15, 25, 35, 45, and 60, 75, 90, 105, 120, 135, and 150 minutes post-infusion. *When this test is used, IR is defined as maintenance of blood glucose concentrations (measured with a hand-held glucometer) above the baseline value for 45 minutes or longer.* The test can be abbreviated to 60 minutes when used in the field, but it is advisable to complete the measurements so that the time taken for the blood concentration to return to baseline can be recorded for future reference. This allows assessment of the response to diet, exercise, or medication.

There is a small risk of hypoglycemia when performing this test, so two 60-mL syringes containing 50% dextrose should be kept on hand and administered if sweating, muscle fasciculations, or profound weakness are observed, or if the blood glucose concentration drops below 40 mg/dL. Note that stress is an important cause of transient IR that can significantly impact CGIT results. In one of our studies, we detected IR in healthy nonobese horses when tests were performed

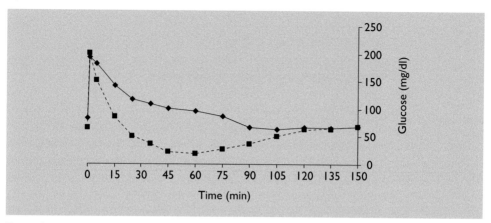

Figure 1. Blood glucose concentrations in an insulin sensitive horse that had returned to baseline by 25 minutes (squares; dashed line) and an insulin resistant horse that remained above baseline until 90 minutes (diamonds; solid line).

immediately after endoscopic examinations (Eiler *et al.*, 2005). Horses must therefore remain calm prior to, and during the procedure to avoid false positive results. Since pain affects results, horses suffering from acute laminitis must be given time to recover before testing is performed.

Dietary management of insulin resistance in horses

Two important questions must be addressed before selecting feeds for insulin resistant horses – is the feed likely to exacerbate IR and will it increase the risk of laminitis?

These questions are related because it is now recognized that pasture-associated laminitis can be triggered by gastrointestinal disturbances arising from alterations in bacterial flora (Van Eps and Pollitt, 2006; Treiber *et al.*, 2006a,b). These disturbances are likely to occur after consumption of lush pasture grass that is rich in fermentable carbohydrate. *It is our hypothesis that IR lowers the threshold for pasture-associated laminitis in horses.* If insulin sensitivity is a determinant of the laminitis threshold, then the goals for managing IR should be to (1) reduce body fat mass in obese animals with the aim of improving insulin sensitivity, (2) avoid feeds that will exacerbate IR, and (3) avoid sudden changes in carbohydrate intake that could alter the intestinal environment and trigger laminitis.

Inducing weight loss in obese insulin resistant horses

Individual horses should be fed according to their metabolic needs. Obese horses that are easy keepers can be placed on a simple diet of hay and a vitamin/mineral supplement. Concentrates are not necessary for these obese horses and weight loss should be promoted by restricting the horse's caloric intake until its ideal weight and body condition have been achieved. This ideal set point differs between individual horses and breeds because the physical stature of the animal varies considerably. The horse must be taken out of an obese state, but it is not necessary for every horse to assume an underweight condition.

Weight loss strategies include dietary management and exercise. Obese horses should be fed enough hay to meet their energy needs, which is usually equivalent to 1.5 to 2.0% of body weight (15 to 20 lbs hay for a 1000-lb horse). Clients should be asked to weigh their hay so that the correct amount is fed. Hay with a low (< 12%) non-structural carbohydrate (NSC) content should be selected for obese insulin resistant horses. Non-structural carbohydrates include simple sugars, starch, and fructans, but NSC measurements reported by commercial laboratories may only include simple sugars (ethanol-extracted soluble carbohydrates) and starches. If the NSC content exceeds 12%, soaking it in cold water for 30 minutes will lower the sugar content prior to feeding. Grass or alfalfa hay can be fed as long as NSC content has been measured. Forages can also be purchased from commercial sources if clients have difficulty acquiring low-NSC hay. Complete feeds and bagged forages are available for insulin resistant horses.

Horses should also receive 1,000 IU vitamin E per day as a supplement because access to green grass has often been restricted. A protein supplement may also be necessary if the quality of hay is poor. Patients that are laminitic should not be exercised until hoof structures have stabilized, but unaffected horses should be exercised regularly. Ideally, horses with EMS should be walked on a lead rope, exercised on a lunge line, or ridden every day.

Avoiding feeds that exacerbate IR

In addition to exercise, care must be taken to avoid feeds that exacerbate IR. The horse with EMS or PPID is similar to a person with diabetes, so excessive sugar should be avoided. Treats containing sugar and sweet feeds should be avoided. Unfortunately, it is very difficult to control sugar intake when horses are grazing freely on pasture. Pasture grass is one of the largest sources of sugar in the horse's diet and the carbohydrate content varies between regions and depends upon soil type, climate, hours of sunlight, and grass species. It also varies according to season and time of day. This creates large fluctuations in carbohydrate intake, which can exacerbate IR and potentially alter the bacterial flora of the large intestine.

Access to pasture must therefore be restricted or eliminated when managing insulin resistant horses and ponies. Sometimes this is only necessary for a few months until weight loss is achieved or PPID becomes better controlled with pergolide therapy. However, there are some insulin resistant horses that have to be permanently housed in dirt paddocks because they are extremely sensitive to changes in pasture grass nutrient content. Thankfully most horses and ponies with PPID or EMS can be managed by limiting grazing time to 1 to 2 hours per day, housing in a grass paddock, strip grazing using an electric fence, or application of a grazing muzzle.

Feeding more calories without exacerbating IR

Some insulin resistant horses have a leaner overall body condition, but exhibit enlarged fat deposits (regional adiposity). Some of these horses are older and may be suffering from PPID. These horses may be exercising strenuously or competing, so they require more calories. If hay is not sufficient to provide these calories, a concentrate must be selected.

Thin insulin resistant horses can be fed concentrates, but care must be taken to provide calories without exacerbating IR. There are three considerations when evaluating feeds for insulin resistant patients: (1) the carbohydrate composition of the feed, (2) the glycemic response that will follow ingestion, and (3) feeding practices. Feeds containing less starch and sugar are appropriate in these situations. It is also advisable to feed hay before concentrates and to feed smaller meals more frequently. Feeding strategies include:

1. A diet consisting of hay with a low (< 12%) NSC content, pelleted specialty feed for IR horses, vitamin and mineral supplement, and 0.5 cup (equal to 125 mL; contains approximately 100g fat) rice bran oil or corn oil twice daily.
2. The same diet with molasses-free beet pulp substituted for pelleted specialty feed.
3. Either of the above diets with rice bran substituted for oil. Rice bran contains approximately 20% fat and 1 lb (approximately 90 g fat) can be fed twice daily.
4. A pelleted specialty feed for geriatric horses in older patients with muscle loss or dental problems (> 20 years of age).

Horses with poorer appetites sometimes refuse to eat beet pulp or specialty feeds, so a small amount of oats may be added to help with this transition. Beet pulp is energy-dense, so it is not an appropriate feed for obese horses, other than as a treat (0.5 cup) to aid in the delivery of supplements. It should be soaked for prior to feeding to remove molasses and simple sugars, and lower the risk of esophageal obstruction. Even molasses-free beet pulp can contain simple sugars, so feed analysis results should be provided by the supplier prior to purchase.

Lowering the risk of laminitis

Laminitis can develop when triggering factors from bacteria within the large intestine enter the circulation after sudden changes in diet (Treiber *et al.*, 2006b; Harman and Ward, 2001). In the case of a grain founder, this occurs when the horse breaks into the feed room and eats too much grain. There is a rapid increase in the amount of sugar arriving in the large intestine which alters the bacterial flora of the colon and causes the intraluminal pH to decrease. This increases intestinal permeability and allows triggering factors to enter the blood. Pasture-associated laminitis develops in a similar way, except that the carbohydrates come from grass consumed on pasture. Laminitis is triggered by changes in bacterial flora, but the individual horse's threshold for laminitis appears to determine whether disease develops.

The decision to allow access to pasture is therefore the most important issue to be addressed when managing EMS. As mentioned previously, there are some horses that are extremely sensitive to alterations in pasture grass composition that must be permanently held off pasture. Owners are strongly discouraged from turning their horse out on pasture if it has suffered from repeated episodes of laminitis and special farrier care is required. However, other horses have a history of laminitis, but the problem has resolved after implementation of recommended weight loss, diet, and exercise programs. These patients can return to limited grazing on pasture. This usually begins with 1 to 2 hours of grazing once or twice a day or turnout with a grazing muzzle.

Pasture grass can also be submitted for analysis and some clients monitor their pastures in order to identify low-risk times of the year for grazing. Basic guidelines for lowering the risk of pasture-associated laminitis include avoiding times when the grass is (1) turning green and growing quickly (spring), (2) first beginning to dry out at the start of a summer drought, (3) rapidly growing after a heavy summer rain, and (4) entering winter dormancy in the fall. In general, the insulin resistant horse or pony should be kept off pasture when the grass is in a dynamic phase. It is also useful to recommend that clients pay attention to their lawn when managing an affected horse; their animal should be held off pasture when the lawn requires more frequent mowing. Horse owners should try to anticipate alterations in grass quality and limit access to pasture at these times.

Treatment of obesity and insulin resistance with levothyroxine sodium

Most horses or ponies with EMS can be effectively managed by controlling the horse's diet, instituting an exercise program, and limiting or eliminating access to pasture. However, there are times when these strategies will not improve the situation fast enough to prevent additional episodes of laminitis. In these situations, drug therapy is warranted to decrease the likelihood of subsequent laminitis episodes that may permanently damage the feet. Weight loss can be accelerated and insulin sensitivity improved by administering levothyroxine sodium (Thyro L®, Lloyd, Inc., Shenandoah, Iowa) in the feed at a dosage of 48 mg /day for 3 to 6 months, which is equivalent to 4 teaspoons (tsp) per day.

We have performed three research studies to evaluate the use of this drug in horses. In our first study, we administered levothyroxine to eight mares according to an incrementally increasing dosing regimen over an 8-week period (Frank *et al.*, 2005). Mares received 24 mg (2 tsp), 48 mg (4 tsp), 72 mg (6 tsp), or 96 mg (8 tsp) for two weeks at a time. Mean body weight decreased and insulin sensitivity increased in treated mares. Our second study evaluated the long-term effects of the drug on body weight and insulin sensitivity in six mares over a 12-month period. Levothyroxine was administered at a dosage of 48 mg (4 tsp) /day and glucose dynamics were measured at 0, 4, 8, and 12 months. Echocardiographic evaluations, complete blood count, and plasma biochemical analyses were also performed at the same times to assess the safety of levothyroxine. Body weight decreased again in response to treatment and this alteration was mirrored by a > 2-fold increase ($P < 0.05$) in mean insulin sensitivity. No adverse health effects were detected. In our final study, we are examining the effects of levothyroxine (48 mg/day) on body weight and insulin sensitivity in horses affected by EMS, and this study is ongoing.

Horses with EMS that we have treated with levothyroxine lose weight and show a reduction in neck circumference. Preliminary results from our current study show that horses (n = 4) on the controlled diet exhibited a 5 cm decrease in mean neck circumference over 6 months, whereas the same measurement decreased by 10 cm in the treated horses (n = 4). We have subjectively observed that the crest becomes softer in treated horses and this finding precedes the reduction in neck circumference. Measured serum total thyroxine (tT_4) concentrations are elevated when levothyroxine is administered at the 48 mg (4 tsp)/day dosage, but these concentrations vary considerably within and between horses. Serum tT4 concentrations range between 40 and 100 ng/mL in treated horses, so levothyroxine is being given at a supraphysiological dosage. However, clinical signs of hyperthyroidism such as sweating or tachycardia have not been observed in treated horses (Harman and Ward, 2001; Ramirez *et al.*, 1998; Alberts *et al.*, 2000).

When levothyroxine treatment is discontinued, horses should be weaned off the drug by lowering the dosage to 24 mg (2 tsp)/day for 2 weeks and then 1 tsp (12 mg)/day for 2 weeks. The benefit of treating horses with levothyroxine at lower dosages for longer periods has not been evaluated scientifically.

References

Al Jassim, R.A., Scott, P.T., Trebbin, A.L., Trott, D. and Pollitt, C.C., 2005. The genetic diversity of lactic acid producing bacteria in the equine gastrointestinal tract. FEMS Microbiol Lett 248:75-81.

Alberts, M.K., McCann, J.P. and Woods, P.R., 2000. Hemithyroidectomy in a horse with confirmed hyperthyroidism. J Am Vet Med Assoc 217:1051-1054, 1009.

Bailey, S.R., Rycroft, A. and Elliott, J., 2002. Production of amines in equine cecal contents in an *in vitro* model of carbohydrate overload. J Anim Sci 80:2656-2662.

Breuhaus, B.A., Refsal, K.R. and Beyerlein, S.L., 2006. Measurement of free thyroxine concentration in horses by equilibrium dialysis. J Vet Intern Med 20:371-376.

Eiler, H., Frank, N., Andrews, F.M, Oliver, J.W. and Fecteau, K.A., 2005. Physiologic assessment of blood glucose homeostasis via combined intravenous glucose and insulin testing in horses. Am J Vet Res 66:1598-1604.

Fowden, A.L., Comline, R.S. and Silver, M., 1984. Insulin secretion and carbohydrate metabolism during pregnancy in the mare. Equine Vet J 16:239-246.

Frank, N., Sommardahl, C.S., Eiler, H., Webb, L.L., Denhart, J.W. and Boston, R.C., 2005. Effects of oral administration of levothyroxine sodium on concentrations of plasma lipids, concentration and composition of very-low-density lipoproteins, and glucose dynamics in healthy adult mares. Am J Vet Res 66:1032-1038.

Frank, N., Elliott, S.B., Brandt, L.E. and Keisler, D.H., 2006. Physical characteristics, blood hormone concentrations, and plasma lipid concentrations in obese horses with insulin resistance. J Am Vet Med Assoc 228:1383-1390.

Geor, R.J., Hinchcliff, K.W., McCutcheon, L.J. and Sams, R.A., 2000. Epinephrine inhibits exogenous glucose utilization in exercising horses. J Appl Physiol 88:1777-1790.

Harman, J. and Ward, M., 2001. The role of nutritional therapy in the treatment of equine Cushing's syndrome and laminitis. Altern Med Rev 6 Suppl:S4-16.

Hoffman, R.M., Wilson, J.A., Kronfeld, D.S., Cooper, W.L., Lawrence, L.A., Sklan, D. and Harris, P.A., 2001. Hydrolyzable carbohydrates in pasture, hay, and horse feeds: direct assay and seasonal variation. J Anim Sci 79:500-506.

Johnson, P.J., 2002. The equine metabolic syndrome peripheral Cushing's syndrome. Vet Clin North Am Equine Pract 18:271-293.

Johnson, P.J., Scotty, N.C., Wiedmeyer, C., Messer, N.T. and Kreeger, J.M., 2005. Diabetes mellitus in a domesticated Spanish mustang. J Am Vet Med Assoc 226:584-588.

Kronfeld, D.S., Treiber, K.H. and Geor, R.J., 2005. Comparison of nonspecific indications and quantitative methods for the assessment of insulin resistance in horses and ponies. J Am Vet Med Assoc 226:712-719.

Mobasheri, A., Critchlow, K., Clegg, P.D., Carter, S.D. and Canessa, C.M., 2004. Chronic equine laminitis is characterised by loss of GLUT1, GLUT4 and ENaC positive laminar keratinocytes. Equine Vet J 36:248-254.

Pass, M.A., Pollitt, S. and Pollitt, C.C., 1998. Decreased glucose metabolism causes separation of hoof lamellae *in vitro*: a trigger for laminitis? Equine Vet J Suppl 133-138.

Patel, J.V., Cummings, D.E., Girod, J.P., Mascarenhas, A.V., Hughes, E.A., Gupta, M., Lip, G.Y.H., Reddy, S. and Brotman, D.J., 2006. Role of metabolically active hormones in the insulin resistance associated with short-term glucocorticoid treatment. J Negat Results Biomed 5:14.

Ramirez, S., McClure, J.J., Moore, R.M., Wolfsheimer, K.J., Gaunt, S.D., Mirza, M.H. and Taylor, W., 1998. Hyperthyroidism associated with a thyroid adenocarcinoma in a 21-year-old gelding. J Vet Intern Med 12:475-477.

Rask-Madsen, C. and King, G.L., 2007. Mechanisms of Disease: endothelial dysfunction in insulin resistance and diabetes. Nat Clin Pract Endocrinol Metab 3:46-56.

Tiley, H.A., Geor, R.J. and Stewart-Hunt, L., 2006. Equine skeletal muscle insulin signaling, gucose metabolism and GLUT4 expression in a dexamethasone model of insulin resistance. J Vet Intern Med (abstract) 20:799.

Treiber, K.H., Kronfeld, D.S., Hess, T.M., Boston, R.C. and Harris, P.A., 2005. Use of proxies and reference quintiles obtained from minimal model analysis for determination of insulin sensitivity and pancreatic beta-cell responsiveness in horses. Am J Vet Res 66:2114-2121.

Treiber, K.H., Kronfeld, D.S. and Geor, R.J., 2006a. Insulin resistance in equids: possible role in laminitis. J Nutr 136:2094S-2098S.

Treiber, K.H., Kronfeld, D.S., Hess, T.M., Byrd, B.M., Splan, R.K. and Staniar, W.B., 2006b. Evaluation of genetic and metabolic predispositions and nutritional risk factors for pasture-associated laminitis in ponies. J Am Vet Med Assoc 228:1538-1545.

Utzschneider, K.M., Carr, D.B., Hull, R.L., Kodama, K., Shofer, J.B., Retzlaff, B.M., Knopp, R.H. and Kahn, S.E., 2004. Impact of intra-abdominal fat and age on insulin sensitivity and beta-cell function. Diabetes 53:2867-2872.

Van Eps, A.W. and Pollitt, C.C., 2006. Equine laminitis induced with oligofructose. Equine Vet J 38:203-208.

Vick, M.M., Sessions, D.R., Murphy, B.A., Kennedy, E.L., Reedy, S.E. and Fitzgerald, B.P., 2006. Obesity is associated with altered metabolic and reproductive activity in the mare: effects of metformin on insulin sensitivity and reproductive cyclicity. Reprod Fertil Dev 18:609-617.

Yki-Jarvinen, H. and Westerbacka, J., 2000. Vascular actions of insulin in obesity. Int J Obes Relat Metab Disord 24 Suppl 2:S25-28.

The application of *Vitex agnus castus* and other medicinal herbs for the symptomatic relief of hyperadrenocoticism and Equine Metabolic Syndrome

Hilary Self
Hilton Herbs LTD, Downclose Farm, North Perrott, Somerset. TA18 7SH
United Kingdom

Introduction

Hyperadrenocorticism is a distressing disease that presents with distinctive symptoms. Diagnosis on the basis of clinical signs and blood tests is supported by carrying out a Dexamethasone Suppression Test. In the equine animal the disease is controlled by conventional medication such as Pergolide mesylate. The progressive nature of the disease, linked with the horse's age usually results in euthanasia on humane grounds.

The presenting signs in both humans and equines are very similar although the aetiology of the disease is not always identical. There would appear to be an increased incidence of the disease in horses although it has been suggested that this is due to improved veterinary techniques of diagnosis, owner awareness of presenting symptoms and increased longevity in horses. The disease has attracted enough interest and subsequent research to warrant the presentation of a paper at the 2002 British Equine Veterinary Congress (Slater, 2002).

Medical herbalists have already identified the dopaminergic action which the herb *Vitex agnus castus* has on the human hypothalamic - pituitary axis, and this action has been confirmed by clinical research (Sliutz *et al.*, 1993).

Brief details of additional herbs which may be useful in this condition are included for reference (Appendix 1).

Orthodox medical understanding of hyperadrenocorticism in the horse and its treatment

'Equine Cushing's disease results from adenomatous hypertrophy of the pars intermedia of the pituitary gland, which produces abnormally high levels of several hormones' (Beech and Garcia, 1985).

The condition of Hyperadrenocorticism in horses, more generally referred to as Equine Cushing's Disease (ECD) has received considerable attention in recent years. The equine press has carried in-depth articles on the disease, resulting in increased awareness of the condition by both the lay horse-owner and the veterinary fraternity. So much so, that at the recent British Equine Veterinary Associations congress held in Glasgow in October 2002 the diagnosis of ECD was presented for discussion (Slater, 2002).

Aetiology of Equine Cushing's Disease

It is now generally agreed, following extensive research into ECD dating from its first description in 1932, that the aetiology of the condition is usually a pituitary adenoma (Pallaske, 1932). This is a benign epithelial tumour derived from the ducts and acini of the pars intermedia of the pituitary gland. The adenoma grows by expansion, has a low mitotic index and does not metastasise. It is rare for an adenoma to develop in the anterior lobe of the pituitary gland. The tumour is often referred to as a pituitary interstitial adenoma (PIA) (Macfarlane *et al.*, 2000).

To appreciate the pathophysiological impact of the disease on the equine animal it is necessary to provide a more detailed understanding of the interaction that occurs between the hypothalamus, pituitary gland and adrenal cortex in healthy animals, which has a direct bearing on the presenting symptoms of ECD in the diseased animal. Horses have evolved as plains animals hunted by carnivores, and as a prey species have developed a keen 'fight or flight' response which is vital to their survival. This is mediated by the autonomic nervous system (ANS) and is under the control of the hypothalamus. The pituitary gland secretes hormones which have specific actions on target organs,these include adrenocorticotrophic hormone (ACTH), which stimulates the adrenal cortex to secrete glucocorticoids, (mainly cortisol). Cortisol is involved in the metabolic response to stress, it increases blood sugar by stimulating the liver to break down protein, making amino acids available for gluconeogenesis. Glucocorticoids also have anti-insulin, anti-inflammatory and anti-allergic actions, and are necessary for the vasoconstrictive action of noradrenaline on blood vessels. The mediation of the stress response by cortisol and adrenaline is reflected in the levels of these hormones in samples of plasma taken from both domesticated and feral horses. Cortisol and corticosterone are the principal plasma glucocorticoids in horses, with ratios ranging from 8:1 in domesticated horses to 168:1 in feral stallions (Bottoms *et al.,* 1972; Kirkpatrick *et al.,* 1977).

As with humans the normal horse exhibits a diurnal rhythm of plasma cortisol with the peak values being found in the morning. It is this fact that supports the use of overnight dexamethasone supression tests as a means of diagnosis of ECD. As will be shown later, ECD horses are resistant to the negative feedback of glucocorticoids and subsequently there is less suppression of plasma cortisol than normal. In one study using this method of diagnosis the test approached almost 100% sensitivity and specificity (Dybdal *et al.,* 1994).

It should be noted that the biological half life of cortisol in the equine animal is approximately 2 hours (Slone *et al.,* 1983).

The pars intermedia of the pituitary gland contains cells that secrete adrenocorticotrophic hormone (ACTH), melanocyte stimulating hormone (MSH), a corticotrophin like intermediate lobe peptide (CLIP) as well as gonadotrophs, thyroid stimulating hormone (TSH), and prolactin positive cells. The hormones produced in the pars intermedia come from the common precursor pro-opiomelanocortin (POMC) which is under the influence of the two neurotransmiters dopamine and serotonin. POMC, ACTH, beta - endorphins, and alpha and beta lipoproteins are also found in the pars distalis area of the pituitary gland, which is under the control of corticotrophin releasing hormone (CRH) from the hypothalamus.

Hypothalamic control can either stimulate or inhibit the secretion of hormones by the pituitary gland via releasing or inhibiting hormones, which are excreted by neurosecretory cells near the optic chiasma. The secretion of pituitary gland hormones is also regulated by negative feedback from the target gland hormones. Negative feedback decreases the secretions of corticotrophs (ACTH), thyrotrophs and gonadotrophs as the levels of their target gland hormones rise (Tortora and Grabowski, 1996).

The negative feedback mechanism

In the healthy animal the '*controlled condition*' exists when the body is in homeostasis. This is the level at which the negative feedback mechanism works, and can be illustrated by the following scenario.

Hypothalamic cells secrete corticotropin releasing hormone (CRH) which stimulates the release of adrenocorticotrophic hormone (ACTH) by corticotrophic cells in the pars intermedia and the pars distalis of the pituitary gland. Increased levels of ACTH in the blood switch on production of glucocorticoids by the adrenal cortex. It should be noted that equine CRH, and ACTH have the same biological activity and amino acid peptide structure as human CRH and ACTH (Livesey *et al.*, 1988; Ng *et al.*, 1981).

In a normal healthy animal, when levels of glucocorticoids (*control*) in the blood fall, there is a reduction in the amount of glucocorticoids bound to the hypothalmic receptor cells that secrete corticotrophin releasing hormone (CRH) (*receptor*). This then triggers (*input*) the secretory cells in the hypothalamus to secrete CRH which then prompts the release of ACTH (*output*) from the pituitary gland. Increased levels of ACTH in the blood stream stimulate the adrenal cortex to produce more glucocorticoids (*effector*) for release into the blood stream (*response*). The increased level of glucocorticoids in the blood then produce the negative feedback effect that inhibits CRH and the subsequent ACTH release.

Horses with ECD have been shown to be highly resistant to the negative feedback effect of glucocorticoids on ACTH secretion. This is because glucocorticoids inhibit secretion of the precurser POMC in the pars distalis, but do not inhibit POMC secretion in the pars intermedia. The

POMC peptides which continue to be secreted by the pars intermedia also potentiate the effect of ACTH. Consequently the ACTH which is secreted has a proportionately greater effect in PIA horses than in normal horses.

Figure 1 illustrates the negative feedback mechanism in both the healthy and diseased animal. Horse plasma contains low levels of cortisol binding globulin. Consequently, even a small increase in total plasma cortisol will result in a large rise of free cortisol. POMC is under the influence of the neurotransmitters dopamine and serotonin and is the precursor to gonadotrophins, TSH, ACTH, MSH, and CLIP. It has been hypothesised that ECD may result from the loss of dopaminergic innervation of the pars intermedia. This hypothesis is supported by research showing that dopamine concentrations in PIA tissue are less than 12% of those in the pituitary gland of normal horses (Millington *et al.*, 1988).

The pathophysiology of the disease arises due to an over production of precursor peptides (POMC), and dysfunction of both the hypothalamus and anterior pituitary caused by the developing pituitary pars intermedia adenoma. A further mechanism which may be implicated in animals with ECD involves Hormone Sensitive Lipase (HSL). In the normal animal, HSL breaks down triglycerides when there is low insulin and low glucose. Ponies and donkeys are often shown to be resistant to the effects of insulin (Jeffcott *et al.*, 1986).

It has been postulated that this has developed as an evolutionary adaptation in ponies and donkeys to enable them to survive in adverse conditions. Should this prove to be true it could explain the higher incidence of ECD reported in breeds such as Welsh Mountain Ponies, other British native pony breeds, and Arabian horses that have evolved to survive in extreme environmental conditions.

N.B. This hypothesis would also account for the increased incidence of Equine Metabolic Syndrome experienced by these animals.

The activity of HSL can also be induced in increased response to stress, triggered by hormones such as ACTH, glucocorticoids, and adrenaline. In stress conditions these hormones override the normal inhibitory action of insulin. This mechanism has the effect of utilising

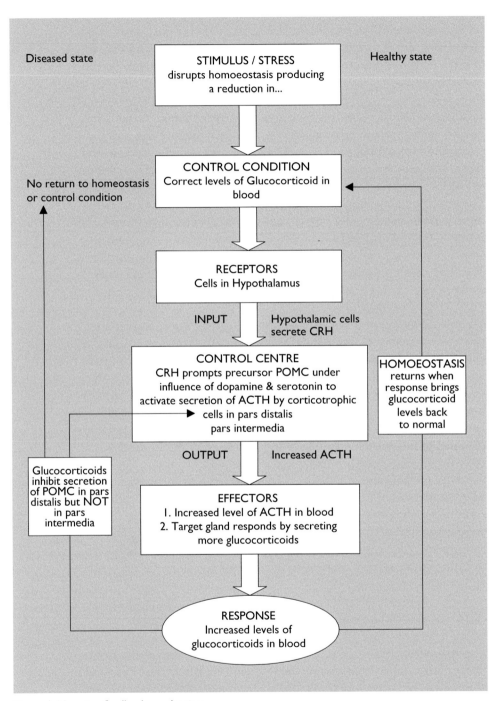

Figure 1. Negative feedback mechanism.

the energy stored in adipose tissue rather than energy from glycogen reserves. Consequently animals with ECD are frequently underweight and are prone to equine hyperlipidaemia which is potentially fatal (Watson, 1998).

This diabetes like condition manifests in ECD symptoms such as:
- glucosuria;
- weight loss;
- polydypsia;
- polyuria;
- retinal dysfunction.

Symptoms of Equine Cushing's Disease

The magnitude of the symptoms of ECD in the horse reflect the importance of the stress response, which has become highly evolved in this prey species to help its survival. ECD is frequently diagnosed in older horses (12 - 35 years).

According to the literature (Hillyer *et al.*, 1992), the following are the most common clinical signs:
- hirsutism;
- weight loss;
- depression/lethargy;
- laminitis;
- polyuria /polydipsia;
- hyperhydrosis.

Other symptoms include:
- bulging supraorbital fat;
- superficial persistent fatty deposits;
- increased susceptibility to infection;
- debility;
- bilateral blindness as a side effect of diabetes;
- infertility;
- spontaneous lactation.

The two most common major complications experienced by ECD horses are diabetes insipidus with subsequent weight loss which occurs in 38% of horses with PIA, and laminitis which may result in euthanasia of the animal. (Van der Kolk, 1993). In addition ECD animals are prone to repeated skin and organ infections. This is thought to be due to the immunosuppressive action of cortisol and the increased levels of blood sugar.

Orthodox treatment

Horses with ECD require symptomatic treatment because arising from their propensity to develop recurrent infections, diabetes insipidus, laminitis, and delayed wound healing.

Conventional treatment modulates the effect of the neurotransmitters dopamine and serotonin. Pergolide mesylate is a type - 2 dopaminergic receptor agonist whose action mimics that of dopamine at the receptors in the pars intermedia. (BNF 43, 2002) The restoration of the normal inhibitory effect of dopamine on the pars intermedia reduces the secretion of prolactin and POMC peptides and subsequent ACTH synthesis and secretion.

It is recommended that the response of the ECD horse is measured carefully in order to establish the minimum dose which is required to control ECD symptoms. Since the drug is replacement therapy for dopamine in the pituitary gland, it must be given for the life of the horse. It is possible that the horse may become resistant to dopamine therapy as is the case in human patients with Parkinson's disease. It is for this reason that the minimum dose necessary is recommended in order to offset the risk of resistance developing.

Cyproheptadine hydrochloride has also been used to treat ECD horses because of its known anti-serotonin effects. Although there have been anecdotal reports of clinical improvement, these findings have not been reproduced by controlled trials. It is not clear at present what precise mechanism of action cyproheptadine may have, nor have any adverse side effects have been reported.

U.S.A. field trial

The only previously reported trial, involving the use of *Vitex agnus castus* for horses suffering with ECD, was conducted using a commercial preparation containing an unspecified quantity of the herb. (Kellon, 2000)

The trial involved 10 horses, all of whom had previously been diagnosed with ECD on the basis of symptoms, or symptoms and blood analysis.

Exact details of the dosage given to each horse are imprecise, however the report states that an 'effective dosage' was found to be 10cc per 200 pounds of body weight, and that 1 litre of the product was sufficient to treat a 1,000 pound (450 Kg) horse for 20 days. This would suggest that a daily dose of 50 mls of product was given to a horse weighing 450 kilos. The concentration of the preparation used is not recorded.

The trial lasted approximately 9 months. The supplementation of the horses with the herbal product was commenced in late winter and continued throughout the summer months into autumn.

Kellon concludes that *'all 10 horses in this field trial showed improved, rapid shedding of abnormal coat as a response to Vitex. Improved energy levels, lifting "depression" and improvement in laminitic pain were noted in all horses that had these symptoms'.*

A drop in blood insulin levels was noted in two of the trial horses. All of them grew back their winter coats. Eight of these appeared to be normal, whilst the remaining two horses grew a classic Cushing's coat. Kellon also observes that these differences in early winter coats would suggest that the hormonal output of pituitary gland tumours or pituitary hyperplasia in horses may be of more than one type, as is described in humans.

The author felt that these trials showed a promising line of research, and decided to investigate these findings further, using a larger number of horses, with a control group in an attempt to establish quantifiable criteria for assessment.

Can *Vitex agnus castus* offer symptomatic relief to ECD horses?

Trial details

Twenty five horses which had been diagnosed with ECD on the basis of clinical signs and blood tests were selected for the trial. The timing of the trial 1.5.2002 - 31.7.2002 was critical. Horses have usually completed their seasonal coat change from winter to summer by the end of April. Therefore any animal that still carried a full winter coat on the 1st of May could reasonably be said to have 'failed to shed'. This is one of the most visual symptoms of ECD and one that often alerts owners to the possibility of the condition. The 25 animals who had all been diagnosed with ECD either by symptoms or blood tests (four horses) were then randomly allocated a number. Animals numbered from 1-15 were placed in the group that would be receiving the *Vitex agnus castus* tincture, whilst animals 16-25 were placed in the placebo group.

Calculation of Vitex agnus castus *dosage for horses*

The recommended daily dosage of *Vitex agnus castus* tincture was calculated on the basis of the human dose. It is recommended that the tincture be given as a single dose each morning, as hormonal regulation is considered to be more receptive at this time (Christie and Walker, 1997).

The animals in the trial group were given a single 10 ml dose of tincture every morning, with food, (10 ml omni mane cum cibos). This dose was irrespective of body weight.

The reasoning behind this regime was two fold:
1. It has been shown that horses like humans, exhibit a diurnal rhythm of plasma cortisol with peak values being found in the morning. This supports the use of the overnight dexamethasone suppression test for diagnosis of Cushings disease in both humans and the equine animal. Having identified this similarity between horses and humans, it was hypothesised that the horse's hormonal regulation may, like humans, also be more receptive in the morning.

In addition it was felt that a single daily dose would result in better owner compliance with the trial criteria.

2. Anecdotal reports suggest that encouraging results had been obtained with a 10 ml dose of *Vitex agnus castus* in ECD horses.

Vitex *agnus castus* - symptomatic relief for ECD horses

Trial results

Results were tabulated on the basis of the most common clinical symptoms as manifested by individual animals, and the degree to which each ECD symptom was affected overall (see Table 1).

Table 1. This tabulates and percentage (%) quantifies the results of the 15 trial horses that received the 10mls of Vitex agnus castus *tincture each morning.*

Symptoms	Pre treatment finding	A % of group	Post treatment findings	B % of PTF	Developed during trial
Hirsutism	15 positive	100.0%	14 improved	93.3%	
Weight loss	10 positive	66.6%	10 improved	100.0%	1 - 6.6%
Lethargy	10 positive	66.6%	9 improved	90.0%	
Laminitis	12 positive	80.0%	9 improved	75.0%	
Polydipsia	9 positive	60.0%	7 improved	77.7%	
Polyuria	9 positive	60.0%	7 improved	77.7%	
Hyperhydrosis	11 positive	73.3%	11 improved	100.0%	
Abnormal oestrus cycle	3 positive	20.0%	3 improved	100.0%	
Muscle wastage	6 positive	40.0%	2 improved	33.3%	
Abdominal distention	7 positive	46.6%	3 improved	42.8%	
Lactation	3 positive	20.0%	3 improved	100.0%	
Abnormal fat deposits	11 positive	73.3%	9 improved	81.0%	

A = Percentage of animals displaying symptoms as percentage of the 15 trial animals.
B = Percentage of improvement displayed by the animals as a percentage of positive.

The following observations are provided in order to expand and elaborate on the tabulated trial results and offer suggestions for the observed changes in the horses receiving the *Vitex* tincture.

Hirsutism

The findings in the trial group receiving the *Vitex* concur with those of the Kellon trial, and represent the most dramatic and highly visible improvements reported by the owners.

The improvement in hirsutism was remarkable. Prior to the trial all of the horses in the trial group receiving the *Vitex* tincture were stated to be hirsute and this was supported by photographic evidence. After the trial fourteen of the fifteen owners reported a resolution of hirsutism in their animals. However in scrutinising the post trial photographs, the author did note that four of the horses had retained a small residue of thicker coat. Several of the owners commented on the improvement in the coat quality, and two owners remarked on the return of normal colour to their ponies coats. It has been noted that animals suffering from ECD have exhibited coat colour change, where black and dark brown hair can change colour to chestnut or light brown (Van der Kolk, 1998).

In the placebo group there was no improvement.

Hirsutism is related to the overactivity of the pituitary corticotrophic cells causing overproduction of ACTH, and overstimulation of the adrenal cortex with subsequent excess of circulating androgens which results in hirsutism. Prolactin has also been linked with abnormal hair coat in ECD horses.

The high percentage of animals showing improvement in the trial group would suggest that *Vitex agnus castus* by acting as a dopaminergic agonist has an effect on the release of CRH. This would concur with Sliutz *et al*'s findings that showed *Vitex* to be capable of blocking the release of prolactin from pituitary gland cells.

Weight loss

Ten (10) of the fifteen (15) trial horses had shown weight loss prior to the trials, and of these ten (100%) showed an apparent increase in weight over the trial period. This particular symptom should not be given too much importance, as many Cushing's horses are kept deliberately underfed, in the belief that this may reduce the risk of developing laminitis. However it may be significant that no such improvement was reported in the placebo group. It has already been shown that in ECD animals the activity of hormone sensitive lipase (HSL) can be induced in the response to stress by hormones such as ACTH, leading to loss of weight. It could be postulated therefore that *Vitex* with its dopaminergic action on the pars intermedia could reduce secretion of POMC peptides and therefore ACTH secretion.

Lethargy

The results support the findings in the Kellon trial.

The questionnaire showed a 66% incidence of animals reported to be lethargic. Whilst this is a difficult symptom to quantify, and as such must be a subjective observation, it can be easily identified in an otherwise energetic animal that usually interacts with its peer group.

The post trial findings were supported by reports of a 90% improvement, demonstrated by increased enthusiasm, energy, mobility and a willingness to interact with the animals peer group. These findings are supported by entries made in the pocket diaries that were provided, for example: *'Today when I turned him out, he trotted across the field towards his friends,…this is something he hasn't done for months'.*

This is in contrast to the 25% improvement reported by the placebo group.

Laminitis

This condition is related to the increase in blood cortisol which has the action of constricting blood vessels. Improvements in the reduction in incidence of laminitis were seen in 75% of the trial animals that had previously been affected. It could be postulated that *Vitex agnus castus* in lowering ACTH ultimately reduces cortisol release and thereby reduces the risk of laminitis.

These findings are in contrast to a 20% improvement in the placebo group.

Laminitis is difficult to predict and despite excellent management can occur without warning. Two horses from the placebo group were euthanased following a recurrence of severe laminitis during the trial period. This developed despite the timely use of conventional medication.

Polydipsia and Polyuria

As previously mentioned, compression of the pars nervosa by the developing pituitary adenoma results in a decrease in the production of antidiuretic hormone (ADH) which is responsible for producing polydypsia and polyuria symptoms.

One of the major complications experienced by ECD animals is diabetes insipidus. Polydipsia and polyuria are typical of the diabetic condition experienced by ECD animals and are attributable to excess corticoids having a glucocorticoid effect. A 77% improvement was reported in the trial group in comparison to 0% improvement in the placebo group. This could confirm the dopaminergic action of *Vitex* on the pars intermedia which reduces the secretion of the precursor POMC and subsequent ACTH secretion which has a direct bearing on the levels of corticoids in the blood.

Hyperhydrosis

There was a 100% improvement in horses which had previously demonstrated excessive sweating. Whilst in part this could be attributed to coat loss (hirsutism improved in 93% of the trial animals), it should be noted that there is a direct effect of ECD on thermoregulation. No such improvement was reported in the placebo group.

Abnormal oestrus cycle

The effects of *Vitex* on the reproductive hormone prolactin are well recognised (Christie and Walker, 1997), *Vitex* also influences the metabolism of androgens in the adrenal cortex. In animals suffering from ECD, reproductive control is impaired, of the three animals that were positive for this symptom before the trial a 100% improvement was reported. The three animals in the placebo group that were positive for this symptom before the trial showed no similar improvement.

Muscle wastage

This symptom showed a 33% improvement. Many of the animals which took part in the trials were aged and had been presenting with ECD symptoms for several years. Muscle wastage is a difficult parameter to assess objectively. It is slow to develop and very difficult to rebuild, especially in animals no longer undertaking active fittening programmes.

Abdominal distention

As in human Cushing's disease this is one of the classic symptoms of the disease. It is directly related to the catabolism of adipose tissue, the redistribution of fat and the slackness of abdominal muscles. It has already been shown that ECD animals have an increased risk of adipose tissue catabolism due to insulin resistance. There was a 46% improvement in the trial group, in comparison with 0% improvement in the placebo group.

Lactation

There was 100% improvement in the three trial group animals that were positive for this symptom, which would confirm the dopaminergic effects that have already been attributed to *Vitex agnus castus*, in the control of hyperprolactinaemia (Sliutz *et al.*, 1993).

Variations in mood

As previously stated the assessment of depression, aggression, or enthusiasm in an animal is subjective and relies on the owner's interpretation of animals behaviour. However there were sufficient remarks in the post trial questionnaire, supported by diary notes to warrant taking these observations into consideration.

In the trial group receiving the *Vitex*, 46% reported an improvement in mood, (lifting of depression), 6% a reduction in aggression, and 20% an increase in energy, enthusiasm and interaction with peer group animals. These comments go to support previous reports of mood improvement by women suffering from PMS (Lauritzen *et al.*, 1997).

Palatability of Vitex agnus castus

Of the fifteen horses receiving the *Vitex*, 53% of owners reported that their horse found the tincture very palatable, 33% stated the horse found it palatable and 13% reported palatability problems and difficulty in dosing. This was a better result than anticipated as the diterpene compounds in *Vitex* impart a bitter taste which it was felt some horses would not find acceptable.

Conclusion

The trial results were extremely encouraging, and showed a reduction or alleviation of symptoms directly related to dopaminergic action on the pituitary, such as lactation, abnormal oestrus cycle, hirsutism, and laminitis. This improvement would seem to support the hypothesis that *Vitex agnus castus* could have a dopaminergic action on the equine pituitary gland.

In humans, *Vitex* is employed extensively for women with PMS and research has confirmed the herb's ability to improve mood. Depression and lack of interest in life was one of the most common comments made by trial animal owners when asked about their horse's mood. Although it is accepted that this assessment is subjective, the improvement in mood reported by the owners of the trial horses receiving the *Vitex* tincture cannot be ignored and may reflect the actions reported in humans.

Dose dependency has been identified in *Vitex*, and this did pose a problem when calculating the dosage for the trial animals. This could be interpreted as a weakness of the trial protocol. Should further work be undertaken, it would be preferable to establish several groups which would receive different doses of tincture and at different times during the day.

To summarise, many aspects of Cushing's disease in horses and humans are similar. For example:

- The function of the hypothalamic - pituitary axis.
- The structure and activity of CRH and ACTH.
- Methods of diagnosis for Cushing's disease.
- Conventional treatment methods.
- The symptoms of Cushing's disease.
- *Vitex agnus castus* has a dopaminergic action on the human pituitary gland and trial results support the hypothesis that these actions are also reproduced in the equine pituitary gland.

In addition this study has demonstrated the following actions of *Vitex agnus castus* in the horse:

- The reduction of hirsutism – with a subsequent reduction in hyperhydrosis.
- The ability to improve energy levels.
- To lift and improve mood.
- The resolution of spontaneous lactation.
- An apparent reduction in the incidence of laminitis.
- A reduction in the diabetic symptoms of polyuria and polydipsia.
- A reduction in abnormal fat deposits.

In conclusion, this trial should encourage further research into the thearapeutic effects of *Vitex agnus castus*. The symptomatic relief demonstrated by these Cushing's disease horses would suggest that this herb could offer exciting potential for the future management of Cushing's disease in humans.

References

Beech, J., Garcia, M.C., 1985. Hormonal responses to TRH in healthy horses and horses with pituitary adenoma. American Journal of Veterinary Research 46: 1841-1843.

Bottoms,G.D.,Roesel, O.F., Rausch, F.D., Atkins, E.L., 1972. Circadian variation in plasma cortisol and corticosterone in pigs and mares. American Journal of Veterinary Research 33: 785-790.

British National Formulary, 2002. 43rd Ed. British Medical Association, Royal Pharmaceutical Society of Great Britain London, U.K.

Christie, S., Walker, A.F., 1997. Vitex agnus castus L: (1) A review of its traditional and modern therapuetic use; (2) Current use from a Survey of Practitioners.

Dybdal, N.O., Hargreaves, K.M., Madigan, J.E., 1994. Diagnostic testing for pituitary pars intermedia dysfunction in horses. Journal of American Veterinary Medical Association. 204: 627-632.

Hillyer, M.H., Taylor, F.G.R., Mair, T.S., Murphy, D., Watson, T.D.G., Love, S., 1992. Diagnosis of hyperadrenocorticism in the horse. Equine Vet. Educ.4: 131-134.

Jeffcott, L.B., Field, J.R., McLean, J.G., O'Dea, K., 1986. Glucose tolerance and insulin sensitivity in ponies and Standardbred horses. Equine Veterinary Journal. 18: 97-101.

Kellon, E., 2000. Herbal offers Hope for Cushing's Syndrome. Horse Journal 7: 3-7.

Kirkpatrick, J.F., Wiesner, L., Baker, C.B., Angle, M., 1977. Diurnal variation of plasma corticosteroids in the wild horse stallion. Comp Biochem Physiol 57A: 179-181.

Lauritzen, C., Reuter, H.D., Repges, R., Bohnert, K.-J., Schmidt, U., 1997. Treatment of premenstrual tension syndrome with Vitex agnus castus. Controlled, double-blind study versus pyridoxine. Phytomedicine, bind 4, nr. 3, 183-189.

Livesey, J.H., Donald, R.A., Irvine, C.H.G., Redekopp, C. Alexander, S.L., 1988. The effects of cortisol corticotropin, alpha-melanocyte-stimulating hormone and AVP secretion in the pituitary venous effluent of the horse. Endocrinology 123: 713-720

Macfarlane, P.S., Reid, R., Callander, R., 2000. Pathology Illustrated,5th Ed. Churchill Livingstone, Edinburgh.

Millington, W.R., Dybdal, N.O., Dawson, R. Jr., Manzini, C., Mueller, G.P., 1988. Equine Cushing's disease: differential regulation of Beta endorphin processing in tumours of the intermediate pututiary. Endocrinology. 123: 1598-1604.

Ng, T.B., Chung, D., Li, C.H., 1981. Isolation and properties of beta-endorphin (1-27), N-acetyl - beta- endorphin, corticotropin, gamma-lipotropin and neurophysin from equine pituitary glands. International Journal Peptide Protein Research.18: 443-450.

Pallaske, G., 1932. Zur kasuistik seltnere Geschwulste bei den Haustieren. Z. Krebsforsch. 36: 342

Slater, J., 2002. Dynamic Endocrine Tests: Do they tell us anything we don't already know? British Equine Veterinary Association 41st Congress, Abstracts of presentations: 64-65.

Sliutz, G., Speiser, P., Schultz, A.M., Spona, J., Zeillinger, R., 1993. Agnus castus extracts inhibit prolactin secretion of rat pituitary cells. Hormone & Metabolic Research.25 (5): 243-284.

Slone, D.E., Ganjam, V.K., Purohit, R.C., Ravis, W.R., 1983. Cortisol disappearance rate and patho- physiologic changes after bilateral adrenalectomy in equids. American Journal of Veterinary Research 44: 276-279.

Tortora G. J., Grabowski S.R., 1996. Principles of Anatomy and Physiology. 8th Ed. Harper Collins, New York.

Van der Kolk, H. 1998. Diseases of the Pituitary Gland, Including hyperadrenocorticism. Watson, T. Metabolic and Endocrine Problems of the Horse. Saunders, W.B. Harcourt Brace and Company Ltd. London.

Van der Kolk, J.H., Kalsbeek, H.C., van Garderen, E., Wensing, Th., Breulink, H. J., 1993. Equine pituitary neoplasia: a clinical report of 21 cases.(1990 - 1992) Veterinary Record 133: 594-597.

Watson,T. 1998. Equine Hyperlipaemia 23-40. Watson, T. Metabolic and Endocrine Problems of the Horse. Saunders,W.B.Harcourt Brace and Company Ltd. London.

Appendix I

In the years following the trials the author formulated a pure herbal mix specifically for animals with ECD. A number of additional medicinal herbs have been selected that have been shown to help alleviate the diabetic symptoms experienced by both ECD animals and animals with other hormonal imbalances such as Equine Metabolic Syndrome.

Details of these herbs are now presented.

Galega officinalis - Goats Rue

Parts used
Whole plant.

Active constituent
Galegine (alkaloid) and Guanidine derivatives.

Actions
Hypoglycaemic–Pharmacological and clinical research in 1930's demonstrated significant hypoglycaemic activity for Galega officinalis, however unlike the biguanide drugs the herb did not have the unpleasant side effects.

Pharmacology
Acts by potentiating the effects of insulin, particularly through the uptake by the cell, the inhibition of intestinal absorption of glucose, the inhibition of gluconeogenesis (particularly significant for ECD and EMS animals). Galega is NOT contraindicated in obese patients. Research on guinea pigs in 1961 found that in some cases the herb had a regenerative effect on the insulin–producing Beta cells of the pancreas[1].

Medicinal uses
Late onset diabetes mellitus (as experienced in ECD and EMS animals)., pancreatitis and digestive problems.

---

[1] Reference: Bone K: Medicherb newsletter. 1989.
Monograph – Galega officinalis – Vicky Ridley MNIMH, CPP 1999.

Dosage
Equine: 3- 4 g of the cut dried herb, twice daily in feed.
Human: Tincture 1:10 45% alc. 3-6 ml daily.

History and folklore
The common name Goats rue arises from the foul smell of the foliage when bruised. The name Galega comes from the Greek gala, 'milk'. *G. officinalis* was once important in the treatment of plague, fevers, and infectious diseases, hence the German name of Pestilenzkraut.

Vaccinium myrtillis - *Bilberry*

Parts used
Fruits and leaves.

Active constituents
Anthocyanins (aka anthocyanosides) these are the blue pigment in the fruit and are structurally related to flavanoids, which are responsible for the vaso-protective and vaso-dilatory action of the plant.
Also contains tannins, pectins and phenolic acids.

Actions
Vasoprotective; antioedema; antioxidant; anti inflammatory; astringent.

Pharmacology
Vascular protective and anti oedema activity.
Anthocyanins improved functional disturbances of fine blood vessels, especially capillaries and were more effective in protecting damaged capillaries than flavonoids and stimulated capillary repair.

Effects on vision
Anthocyanins have demonstrated an affinity for the pigment epithelium of the retina *in vitro*, Biberry hastens the regeneration of rhodopsin *in vitro* and *in vivo* after injection. Rhodopsin is a light sensitive pigment found in the rods of the retina, it must be quickly regenerated in order to maintain visual sensitivity. Bilberry helps accelerate adaptation to the dark and improved night vision in open trials conducted with air traffic controllers, pilots and car drivers.

Antioxidant activity

In vitro antioxidant activity was noted in a number of models, including scavenging of super oxide anions and inhibition of lipid perioxidation.

Wound healing activity

Bilberry extract accelerated the process of spontaneous healing of experimental wounds after topical application.

Anti platelet activity

Bilberry extract demonstrated strong anti–platelet activity *in vitro*, prolonged bleeding time without affecting blood coagulation *in vivo*.

Medicinal uses

Recommended uses confirmed and established by clinical trials:
Peripheral vascular disorders of various origins, including Raynauds syndrome, venous insufficiency especially of the lower limbs, symptoms caused by decreased capillary resistance; conditions involving increased capillary fragility such as diabetic and hypertensive retinopathies, vision disorders including altered micro circulation of the retina.

Dosage

Equine: 3-4 g of dried Bilberry fruit twice daily in feed.
Human: Fluid extract 1:1. 3 – 6 ml per day.

N.B. It should be noted that Vaccinium leaves have a role to play in the treatment of late onset diabetes, so certainly warrant investigation in this area, they contain glucoquinones which may have a role in reducing blood sugar levels.

Most pharmacological information derived from:
Mediherb Professional review by K. Bone, no 59. 1997.
Monograph – Vaccinium myrtillus L. – Vicky Ridley MNIMH, CPP 1999.

Cynara scolymus – Globe Artichoke

Parts used
Flower heads, leaves.

Active constituents
0.5 – 4.5% of bitter sesquiterpine lactones in the leaves, the principle one being cynaropicrin.
Flavonoids found in the leaves are mainly derivatives of luteolin. Cynara also contains coumarins and plant phenolic acid derivatives.
The plant phenolic acids and their derivatives found in Cynara include cynarin and cryptochlorogenic acid, neochlorogenic, chlorogenic and 1,3–dicaffeylquinic acids.

Actions
Bitter tonic, choleretic, cholagogue, diuretic, hepatoprotective and hepatic trophorestorative.
Reduces blood lipids, serum cholesterol and blood sugar. Due to the combination of choleretic and diuretic actions this plant also makes a good depurative.

Pharmacology
All the phenolic acids are potentially unstable, but careful drying of the leaves will insure that losses are minimal.
Plant phenolic acids probably function as anti oxidants and may act as co factors in several enzyme systems.
In one clinical study 60 patients with elevated serum lipids were treated for 50 days with cynarin (500 mg) or placebo. Total cholesterol was lowered significantly by about 20%. In addition there was also a significant reduction in average body weight for the treatment group of about 5 kg.

Medicinal uses
For chronic liver diseases, jaundice, hepatitis, sluggish digestion.
To lower serum cholesterol and triglycerides, especially in obese individuals.
As a depurative for rheumatic conditions, osteoarthritis, chronic skin diseases.
Renal insufficiency.

Dosage
Equine: 5 grms twice daily in feed.
Human: Tincture 1:3. 5 ml three times daily.

Monograph – Cynara scolymus – Vicky Ridley MNIMH, CPP. 1999.

Silybum marianum - *Milk Thistle*

Parts used
Seed.

Active constituents
Principally Silymarin which is a mixture of flavanolignans.
Fixed oil with large amounts of linoleic acid (60%), Oleic acid (30%) and palmitic acid (9%), in the form of triglycerides, tocopherol, sterols, cholesterol.

Pharmacolog
Silymarin is one of the most intensively investigated plant substance.
The main activity is as a hepatoprotective and anti–oxidant.
It's effectivity as a hepatoprotective has been demonstrated in numerous experiments and clinical trials with poisons such as ethylalcohol, tetrachloride, frog virus, and poisons of the amanita mushroom.
Silymarin acts in several ways. Firstly acting as a membrane stabiliser and protector, thought to be brought about through an antioxidant and radical scavenging action.
Silymarin also interacts with the cells of the liver membrane by blocking binding cites and hindering the uptake of toxins. The anti oxidant activity is 10 times that of Vitamin E. Silybin also increases the synthesis of ribonucleic acids, this occurs by stimulating the enzyme RNA polymerase A. Thus increasing the regenerative capacity of the liver through the developlment of new cells.
Trials have shown that the prophylactic use of Silybum can inhibit industrial chemical, pharmaceutical and alcohol induced liver damage and speed up the restoration of impaired liver damage.

Medicinal uses
Hepatoprotective effect against liver poisons such as pharmaceuticals, alcohol, amanitin, phalloidin.

Chronic inflammatory liver disorders such as cirrhosis, hepatitis, fatty liver.

A tonic for the whole digestive system, hepato–tonic and as a bitter. Although the therapeutic spectrum of Silybum may appear rather narrow, due to the enormous importance of the liver as an organ, milk thistle is a very important tool in the herbalist repertoire.

Dosage

Equine: 4 –5 g of bruised milk thistle seed twice daily in feed.
Human: Tincture 1:5 25% alc. Up to 6 ml three times a day.

Monograph - Silybum marianum – H.H. Zeylstra 1998.

Gymnema sylvestra – *Small Indian Ipecac*

Parts used
Leaves.

Active constituents
Gymnemic acids – a chemically complex mixture of saponins.

Actions
Hypoglycaemic, antidiabetic, hypocholesterolaemic.

Pharmacology
The molecular structure of gymnemic acids are similar to that of glucose molecules. On tasting Gymnema the molecules fill the receptor locations on the taste buds for a period of 1 – 2 hours therby preventing the taste buds from being activated by any sugar molecules present in the food. This effect also appears to occur at the receptor sites in the small intestine, therby preventing the intestine from absorbing glucose. The leaves are also thought to increase insulin secretion.
Studies as early as 1930 conducted in India showed that the leaves caused hypoglycaemia in experimental animals. This hypoglycaemic action is explained on the assumption that the drug directly stimulates the insulin production of the pancreas, since it has not direct effect on carbohydrate metabolism. Early studies showed no significant effect on blood sugar levels in animals with normal blood sugar profiles but there was significant reduction in animals with hyperglycaemia.

Hilary Self

Gymnema extract and gymnemic acid significantly depressed the portal release of gastric inhibitory peptide after intraduodenal glucose infusion. They may be interacting with a glucose receptor that exists for the release of gastric inhibitory peptide.

Clinical studies
1. Two long term studies (without placebo controls) have shown promising results. The first done on insulin–dependant diabetes mellitus found the Gymnema extract reduced insulin reequirments and fasting blood glucose levels. There was some suggestion of enhancement of endogenous insulin production, possibly by pancreatic regeneration.
   The second study was done on non–insulin dependant diabetics and produced similar results and hypoglycaemic drug requirements were reduced. Fasting and post–prandial serum insulin levels were elevated in the Gymnema group compared to controls taking conventional drugs.
2. Clinical research under double–blind conditions found that gymnemic acid considerably diminished the sweet taste. The study also revealed that gymnemic acid also significantly decreased appetite for up to 90 minutes after the sweet–numbing effect.

Medicinal uses
Some cases of diabetes will respond quickly, but best results come after 6 – 12 months of continual use.
Because Gymnema anaesthetises the sweet taste buds and the appetite for some time, it may have implications in weight control.

Note: Caution must be the key word when using any antidiabetic herb, the individual should always be monitored.

Dosage
Equine: 1 ml twice daily in food.
Human: 5 – 10 ml / day of a 1:1 extract for diabetes.

Less may be needed if combined with other anti diabetic herbs, or if on orthodox medication.
1 – 2 ml /day of a 1:1 extract for sweet craving and sweet taste depression. In the latter case the drops should be applied directly to

the tongue and rinsed off after 1 minute. This can be done at 2 – 3 hour intervals.

History and folklore

The safety and efficacy of Gymnema sylvestre has been demonstrated by the fact that it has been used for more than 2,000years in traditional Ayurvedic medicine. It came to be known as 'destroyer of sugar' because in ancient times Ayurvedic physicians observed that chewing a few leaves depressed the taste of sugar.

Monograph on Gymnema sylvestra – Vicky Ridley MNIMH, CPP. 2000.
Bibliography: Bone K, (1997) Clinical applications of Ayurvedic and Chinese herbs. Phytotherapy press.

Food-allergy in horses

Regina Wagner[1], Derek C. Knottenbelt[2] and Birgit Hunsinger[3]
[1]Vetderm-Service, Austria
[2]University of Liverpool, United Kingdom
[3]Laboklin GmbH&Co.KG, Bad Kissingen, Germany

Definitions

Food-hypersensitivity (FH), food-intolerance, abnormal reaction to food (Ackerman, 1993) or allergic skin-reaction to food-components (Peters, 1997) are synonymously used for food-allergy.

Food-allergy (allergic / hypersenstitivity responses to dietary components) is the most frequent skin disease in dogs and cats following flea-allergy and atopic dermatitis (allergic reaction to environmental allergens) (Scott et al., 1995).

Skin disease due to allergic reactions to food-components are also recognised in equine practice (Hutyra and Marek, 1922; Dietz and Wiesner, 1982; Francqueville and Sabbah, 1999; Volland-Francqueville and Sabbah, 2004). The reactions can lead to considerable discomfort due to pruritus, skin lesions and secondary infections, which altogether can result in a reduction of the value in usage of the horse.

Since there are few publications concerning food-allergies in horses, the following general background is based on publications of the small animal and human medicine. However, it can be assumed, that in large part the findings can be applied to horses, too.

Food-allergy is recognised as a non-seasonal often pruritic skin disease caused by the ingestion of a substance found in the horse´s diet. Suggested allergenic agents include proteins, carbohydrates, and also preservatives or antioxidants.

Several different approaches to the clinical problem have been identified (Figure 1):

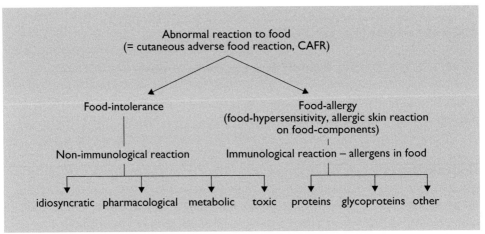

Figure 1. Classification of food-intolerance.

1. Food-allergy is a form of the adverse food reaction, where an abnormal immunological reaction occurs (Halliwell, 1992).
2. Cutaneous adverse food reaction marks a clinically abnormal reaction to a food-component without reference to the aetiology. Often, however, the term food-allergy or -hypersensitivity is used without the immunological basis being confirmed.
3. Food-intolerance is an abnormal reaction to food or food-components without immunological basis. The intolerance can be idiosyncratic, metabolic, pharmacological or toxic origin.
    a. Food- idiosyncrasy is a quantitatively abnormal, though not immunological reaction, that imitates an allergy (Halliwell, 1992). It is caused by a non-specific mast cell degranulation due to 'histamine-releasing-factors' or by histamine-rich food. The latter is less relevant in horses (although possibly, some horses might have access to strawberries, tomatoes,...).
    b. The pharmacological food-reaction produces a drug-like effect in the organism, e.g. methylxanthine in chocolate.
    c. The metabolic food-reaction results from an effect in the metabolism of the recipient, for example a primary gastrointestinal disease (Ackerman, 1993) or the deficiency of the enzyme lactase.
    d. Toxic reaction (food-poisoning) is caused by toxins, e.g. mycotoxins or the botulism-toxin, contained in the food or being produced by micro-organisms in the food.

Since the pathological mechanisms are not sufficiently known, the working group of the ACVD (American college of Veterinary Dermatology) recommends the term 'cutaneous adverse food reaction (CAFR)' as suitable independent from the pathogenesis (Hillier and Griffin, 2001).

Pathogenesis

It is generally assumed that food-allergy is an IgE-mediated reaction. Absorption of food-allergens leads to specific IgE-antibodies being bound at the surface of mast cells and basophiles, causing an allergy of the immediate and delayed type.

It is hypothesised that beside type 1 reaction (immediate type), allergy type 3 (Immune complex type) and 4 (delayed or cell-mediated type) are involved, too. The allergen is frequently a glycoprotein, that induces allergic effects by heating, preparation or by digestion (Scott *et al.*, 1995). The food-intolerance generally is caused by substances (proteins, specific plant related antigens), causing a non-immunological mast cell degranulation (Ackerman, 1993) or in some cases by vasoactive amines (such as are present in cheese, chocolate, coffee, bananas, etc.).

Food-hypersensitivity and food-intolerance are not distinguishable clinically (Ackerman, 1993). No limitation of age could be observed in dogs and cats exhibiting the problem, e.g. it can affect very young or very old animals (Rosser, 1993; Rosser, 1996; Wagner and Horvath, 1999). Frequently, the disease occurs when the animals are fed the same food for a long time. In that case it is often difficult to convince the owner of the need to alter the diet because the patient tolerated the food well previously.

Clinical symptoms in horses

The symptoms of food-allergy are variable and often it is difficult to distinguish from other allergies or other pruritic diseases. Scott and Miller (2003) classified food hypersensitivity in horses as:
1. Urticaria (pruritic or not).
2. Angioedema, multifocal or generalized pruritus, often in the perineal and tail region.

Commonly affected areas include face, neck, trunk, and rump. The main differentials include atopic dermatitis (allergy to environmental allergens), contact and insect-hypersensitivity, pediculosis, oxyuriasis, and chorioptic and psoroptic mange.

'Urticaria' is an end stage (final common pathway) reaction-pattern, not a disease! It can have various causes, but among other things it is a symptom of atopic dermatitis, food- or insect-hypersensitivity.

Clinical symptoms range from mild pruritus +/- urticaria to self-induced excoriations, secondary infections and secondary lesions, like alopecia, hyperpigmentation and others. Besides skin-reactions, however also latent enteritides can occur (Francqueville and Sabbah, 1999).

Diagnosis

The final diagnosis of the allergies and identification of relevant allergens is the basic requirement for a promising therapy with long lasting effect. A careful and precise history represents the central component of the diagnosis. The second main pillar is a clinical-dermatological examination. This leads to a list of differential-diagnoses, which are substantiated or excluded with help of suitably selected diagnostic tests.

An initial presumptive diagnosis may result from a detailed history (season, attitude, feeding, etc.), the clinical examination and the responses to treatment either by the owner of previous veterinary prescriptions.

Further examination for ectoparasites and endoparasites (e.g. lice, if not found with diagnostic therapy; Chorioptes, Psoroptes, and Oxyuris equi infestation and specifically at the ventral center line -Onchocerca dermatitis), as well as a bacteriological and mycological examination of a skin scraping (dermatophilosis and dermatophytosis – fungal culture) should be undertaken primarily to eliminate alternative causes of the clinical problem.

A biopsy may be helpful for confirming the suspicion of an allergic dermatitis, but this seldom provides either specific pathological or aetiological information.

The gold standard of diagnosis for food-hypersensitivity is an elimination-diets lasting up to 12 weeks with subsequent provocation diet. The novel protein diet must be individualized for each patient, on the basis of careful dietary history. Very simple avoidance of treats or the various equine additives brings often relief. The object of the dietary test is to feed something that is not normally fed, to feed a food free of additives, not commercial food, but just pure grain and to feed only one source of hay or grass and if at all necessary one kind of grain without anything else. Also no carrots, apples, bread or not at all commercial treats.

Elimination diets require to be prolonged - at least 2 months, often 12 weeks is advised because in a study with dogs Rosser (1996) showed that improvement of the symptoms does not occur with shorter duration. Previously 3 weeks of elimination diet were recommended, but only 26% of allergic animals are identified within that time span. A single exposure to an antigen can release a group of cytokines (histamine-releasing factors), which can continue the histamine-release even without subsequent antigen contact. Therefore responses to an elimination diet can take a long time. Furthermore, repeated or continuous adverse reaction to food causes a loss of the intestinal villi and the recovery from this alone may take up to 8-10 weeks (Halliwell, 1992).

The suspicion of an adverse reaction to food is confirmed when the old diet is reintroduced and clinical signs appear after 7 hours to 14 days. The elimination diet is then restarted and after remission one new substance at a time is introduced in order to find out the specific substance(s) involved. Horses with concurrent atopic dermatitis or insect hypersensitivity may only partly improve.

Therapy

The only current therapy is avoidance of the offending agent. If the allergen is known, and it was obviously fed, that can be a very convenient therapy. But if the allergen is a 'hidden' allergen, e.g. soy which can be included as binding substance in foods, without being labelled, avoidance can be difficult. Normally such horses must not eat any 'treats' and no commercial food, but should be fed with just simple grass or hay and organic, natural grain.

Regina Wagner, Derek K. Knottenbelt and Birgit Hunsinger

References

Ackerman, L., 1993. Adverse Reaction to Foods. J Vet Allergy Clin Immunol. 1 (1): 18-22.

Dietz, O. and Wiesner, E., 1982. Handbook of Horse Illnesses, Meager, Basel.

Evan, A.G., Paradis, M.R. and O'Callaghan, M., 1992. Intradermal testing of horses with chronic obstructive pulmonary disease and recurrent urticaria.. J. Vet. Res. 53 (2): 203-208.

Francqueville, M. and Sabbah, A., 1999. Chronic urticaria in sports horses. Immunol. 31: 212-213.

Halliwell, R.E.W., 1992. Comparative aspects of food intolerance. Vet Med. 87: 893-899.

Hillier, A. and Griffin, C.A., 2001. The ACVD task force on canine atopic dermatitis (X): ist there a relationship between canine atopic dermatitis and cutaneous adverse food reactions? Vet Immunol Immunop. 81: 227-231.

Hutyra, F. and Marek, J., 1922. Particular pathology and therapy of the pets. Volume 3, 6. Edition. Fishers, Jena: pp. 534 and following.

Kolm-Stark, G. and Wagner, R., 2002. Intradermal skin testing of Icelandic horses in Austria. Equine Vet. J. 34 (4): 405-410.

Lebis, D., Bourdeau, P. and Marzin-Keller, F., 2002. Intradermal skin tests for equine dermatology: a study of 83 horses. Equine vet. J. 34: 666-672.

Panhuizen, M.R., Koeman, J.P. and Sloet van Oldruitenborgh-Oosterbaan, M., 2003. Intradermaler skin-test: Relationship between macroscopic skin-reactions and histological judgment. XV. Convention about horse-illnesses, 14.-15. March 2003, meal.

Peters, S., 1997. Futtermittelallergie. Proc. Dermatologieseminar 2 AKVD/VÖK, 10/97, Vienna: 1-8.

Rosser, E.J. Jr., 1993. Diagnosis of food allergy in dogs. J Am Vet Med Assoc. 203: 259.

Rosser, E.J. Jr., 1996. Diagnosis and treatment of food allergy in dogs and cats. 12th Proc. Ann. Memb. Meet. AAVD/ACVD, Las Vegas, Nevada.

Scott, D.W., Miller, E.W. and Griffin, C., 1995. Immunologic Skin Diseases. In: Small Animal Dermatology, Philadelphia W.B. Saunders pp485 -613.

Scott, D.W. and Miller, W.H., 2003. Equine Dermatology. Saunders, USA.

Volland-Francqueville, M. and Sabbah, A., 2004. Recurrent or chronic urticaria in thoroughbred racehorses:. clinical surveillance. Universe-erg. Immunol. (Paris, 36: 9-12.

Wagner, R. and Horvath, C., 1999. Capelin & Tapioca dry food in dogs and cats with food allergy. Proc. Ann. Memb. Meet. AAVD/ACVD, Maui, Hawaii.

What's new in equine sports nutrition (2005-06)?

Raymond J. Geor
Middleburg Agricultural Research and Extension (MARE) Center, Virginia
Polytechnic and State University, Middleburg, Virginia 20117 USA
rgeor@vt.edu

Horses are used for a wide variety of athletic pursuits and exercise physiology and sports medicine continue to be important areas of equine research. In this paper, I review three recent studies concerning muscle glycogen replenishment in horses after exercise, an area of interest given the importance of this fuel store to exercise performance and the knowledge that the muscle glycogen synthesis rate after exercise in horses is much slower when compared to humans and rodents. Two of the studies reviewed were presented at the 7th International Conference on Equine Exercise Physiology, Fontainebleau, France in 2006.

Jose-Cunilleras, E., Hinchcliff, K.W., Lacombe, V.A., Sams, R.A., Kohn, C.W., Taylor, L.E. and Devor, S.T., 2006. Ingestion of starch-rich meals after exercise increases glucose kinetics but fails to enhance muscle glycogen replenishment in horses. The Veterinary Journal 171: 468-477.

Objectives

1. To determine the effect of glucose supply (post-exercise ingestion of starch-rich meals vs. fibre-rich meals vs. withholding feed for 8 hours) on stable-isotope tracer-determined whole-body glucose rates of appearance and disappearance and on muscle glycogen replenishment during an 8 hour period after glycogen-depleting exercise.
2. To determine the effects of different caloric intakes on muscle glycogen replenishment over a 24 hour period after exercise. It was hypothesized that ingestion of starch-rich meals during the hours

following exercise would result in increased glucose appearance and uptake, and enhanced glucose deposition as muscle glycogen.

Background

Glycogen stored within muscle is an important fuel source during exercise. In horses, previous studies have shown that low muscle glycogen content prior to exercise results in decreased athletic performance during moderate and high-intensity exercise. Furthermore, postexercise muscle glycogen synthesis appears to be quite slow in the horse, with as much as 72 h required for complete replenishment of glycogen stores. Inadequate muscle glycogen replenishment could adversely affect the performance of horses participating in several events in one day or on consecutive days. Therefore, there is interest in the development of nutritional strategies that optimize muscle glycogen synthesis in horses after exercise. This paper is one of several recent studies investigating the mechanisms of muscle glycogen replenishment and/or the effects of different dietary treatments on the rate of muscle glycogen synthesis.

Overview of the study

Seven moderately trained Thoroughbred horses were used in a three-way crossover design. The horses were subjected to three consecutive days of strenuous treadmill exercise and feed restriction (8.5 kg of mixed grass hay), a protocol designed to lower muscle glycogen content by at least 55% of the initial value. The daily exercise protocol comprised a warm-up, followed by running at a speed equivalent to 70% of maximal oxygen uptake (VO_{2max}) for 15 min, and then 5 min at 90% VO_{2max}. After a 30 min rest, horses completed six 1 min sprints at 100% VO_{2max}, with a 5 min rest in between. Feed was withheld for 18 h prior to the third consecutive day of glycogen-depleting exercise. During the 8 h period after this last bout of exercise, horses were either: (1) offered two meals of cracked corn (C) (2.2 ± 0.2 kg/meal [average ± SD], ~7.4 Mcal digestible energy [DE] per meal; grain trial); (2) offered two isocaloric meals of grass and alfalfa hay (H) (3.4 ± 0.4 kg/meal, ~7.4 Mcal or ~30 MJ DE/meal; hay trial); or (3) not fed (NF). At 8 h after exercise, all horses were given another half of the daily DE requirements in the form of mixed grass and alfalfa hay (6.9 ± 0.8 kg; ~14.8 Mcal DE). Whole body glucose kinetics (using a stable isotope tracer of glucose) was studied during and for 8 h after exercise, and muscle biopsies were taken at intervals over 24 h after exercise for measurement of muscle glycogen content. The trials were separated by two weeks, and the order of trials was randomized for individual horses.

Main findings

In all trials, muscle glycogen content was decreased by ~60-65% after three days of strenuous treadmill exercise. There was no statistical difference between hay-fed and corn-fed horses in the amount of DE ingested over the 8 h postexercise period (~10 and ~13 Mcal DE for H and C, respectively). The estimated amount of starch ingested over 8 h after exercise was more than 14-fold higher in corn-fed than in hay-fed horses (5.3 ± 0.4 vs. 0.38 ± 0.03 g starch/kg body weight). As expected, ingestion of corn meals rich in starch, when compared to feed withholding, resulted in moderate hyperglycemia (5.7 ± 0.3 vs. 4.7 ± 0.3 mmol/L) and hyperinsulinemia (79.9 ± 9.3 vs. 39.0 ± 1.9 pM). The tracer-determined estimate of whole-body glucose utilization during the 8 h period after exercise was ~3-fold greater in corn-fed horses when compared to NF. However, despite this increase in glucose turnover muscle glycogen content did not differ between treatments at 8 and 24 h after exercise, and muscle glycogen stores remained ~40-45% below pre-exercise levels at 24 h.

Practical interest

Consistent with previous studies by this group (Lacombe *et al.* 2004) and others (e.g. Davie *et al.*, 1994), the present study has demonstrated that feeding status does not affect muscle glycogen storage during the first 24 h after exercise-induced muscle glycogen depletion. In a previous study (Lacombe *et al.*, 2004) performed by this group at Ohio State University, a high starch diet (75% grain, 25% hay) provided over a 72 h period after the identical glycogen-depleting exercise protocol did enhance muscle glycogen replenishment, but the increase in glycogen content was not evident until 48 h and glycogen stores did not return to pre-exercise levels until 72 h. In humans, glycogen replenishment after exercise is maximized when carbohydrate supplements are ingested at a rate of 1.0-1.5 g/kg body weight immediately after exercise and every 2 h for up to 6 h, i.e. a total of 4 to 6 g carbohydrate/kg (Ivy,

1998), an amount similar to that ingested by the corn fed horses of the present study. It is possible that, in horses compared with humans, the more limited capacity for digestion and absorption of starch and other sources of glucose in the small intestine may constrain the rate of muscle glycogen replenishment due to lack of glucose availability in skeletal muscle. This possibility was examined in the study by Geor *et al.* (2006) described below.

Comments on the study

The authors employed elegant techniques, in particular tracer-determined glucose kinetic measures, to evaluate the effects of common feedstuffs on glucose turnover and muscle glycogen storage after glycogen-depleting exercise. Some may question the relevance of the exercise protocol used to deplete muscle glycogen content relative to real world competition exercise (3 consecutive days of strenuous exercise vs. e.g. the cross-country phase of a 3-day event). Nonetheless, overall this study has contributed to our understanding of factors limiting muscle glycogen replenishment in horses after exercise.

What next?

Clearly, more work is need to better understand the mechanisms underlying the relatively slower muscle glycogen synthesis rate after exercise in horses when compared to humans and other species. For example, future studies should examine the fate of ingested glucose (oxidation vs. storage as glycogen in liver or muscle).

Geor, R.J., Larsen, L., Waterfall, H.L., Stewart-Hunt, L. and McCutcheon, L.J., 2006. Route of carbohydrate administration affects early post exercise muscle glycogen storage in horses. Equine Veterinary Journal Supplement 36: 590-595.

Objective

To compare the effects of oral and intravenous (IV) glucose administration on muscle glycogen storage in horses after glycogen-depleting exercise. It was hypothesized that glucose delivery from the gastrointestinal tract limits the rate of muscle glycogen storage in horses. Therefore, it was predicted that muscle glycogen storage would be enhanced with IV glucose administration but not with an equivalent oral dose.

Background

As mentioned, glycogen replenishment after exercise in humans is greatly accelerated when between 4 and 6 g carbohydrate/kg body weight is ingested during a 6 h period following exercise (Ivy, 1998). In contrast, oral administration of glucose or a glucose polymer at 2-3 g/kg immediately after exercise in horses does not result in enhanced muscle glycogen replenishment (Davie *et al.*, 1994; Nout *et al.*, 2003). On the other hand, IV infusion of glucose (6 g/kg body weight over an 8-9 h period) resulted in almost complete replenishment following exercise that depleted glycogen to 50% of its initial value (Davie *et al.*, 1995; Lacombe *et al.*, 2001). These observations demonstrate a marked difference in the rate of post exercise muscle glycogen storage associated with the mode of carbohydrate delivery and suggest that glucose delivery from the gastrointestinal tract may be one factor that limits the rate of glycogen storage. However, no study has directly compared the effects of equivalent dosage of glucose administered by the oral and IV routes.

Overview of the study

A randomized crossover design with 3 glucose supplementation treatments (control, oral and IV) was used. Seven fit Standardbred horses completed each of 3 trials comprising a 120 min period of treadmill exercise designed to deplete muscle glycogen by ~50% of its initial value and a 6 h post exercise treatment period. After completion of the glycogen-depleting exercise, horses received: (1) an IV glucose infusion (IV; 0.5 g/kg bwt/h for 6 h), (2) oral glucose boluses (OR; 1 g/kg bwt at 0, 2 and 4 h post exercise), or (3) no glucose supplementation (CON). Feed was withheld for 10 h before the exercise protocol, which consisted of running on an inclined (3°) treadmill with a 5 min warmup at 4 m/s, 30 min at a speed equivalent to 50% of VO_{2max}, 15 min at 75% VO_{2max}, and 5 min at 90% VO_{2max}. After a 15 min rest, the horses then completed 6 sprints (1 min/sprint) at a rate calculated to achieve 100% VO$_{2max}$, with a 5 min walk between each sprint. Blood samples for measurement of glucose and insulin concentrations were collected before exercise and at intervals during the 6 h treatment period. Muscle biopsies for measurement of muscle glycogen content and glycogen synthase activity were taken before and after exercise and at 3 and 6 h of treatment.

Main findings

Mean plasma glucose concentrations were significantly higher in IV and OR than in CON throughout treatment; overall average concentration (± SEM) during the 6 h period in CON, OR and IV were, respectively, 4.4 ± 0.5, 6.5 ± 0.9 and 15.4 ± 3.0 mM. The average insulin response for IV (75.3 ± 19.2 μU/ml) was significantly greater when compared to OR (35.2 ± 10.1 μU/ml), while the insulin response in CON (7.2 ± 2.1 μU/ml) was significantly lower when compared to IV and OR. In each treatment, muscle glycogen content was decreased by ~52% from initial (pre-exercise) values. The overall mean rate of glycogen storage during the 6 h treatment was more than 2-fold higher in IV (20.9 ± 7.3 mmol/kg/h dry muscle [dm]) than in CON (6.9 ± 3.7 mmol/kg/h dm) and OR (9.3 ± 4.9 mmol/kg/h dm). As a result, muscle glycogen content was significantly higher in IV than in CON and OR at 6 h of treatment. In IV, muscle glycogen at 6 h of recovery was 75.7 ± 5.1% of the pre-exercise value, whereas glycogen content in CON and OR was, respectively, 51.4 ± 4.3% and 56.5 ± 6.2% of the initial value. Glycogen synthase activity was significantly higher in IV when compared to the other treatments at 3 h of recovery.

Practical interest

Muscle glycogen storage in horses during a 6 h period after glycogen-depleting exercise was enhanced by IV glucose administration (3 g/kg) but not an equivalent glucose dose delivered via the gastrointestinal tract. While these results support a difference in glycogen storage rate based on the route of glucose administration, the findings were not wholly consistent with the working hypothesis because the level of hyperglycemia and hyperinsulinemia achieved during the oral treatment has been shown to accelerate glycogen storage in other species, including humans. On a practical level, however, it is clear that glucose supplementation via the IV route should be considered when rapid replenishment of muscle glycogen stores is desired; for example, after completion of the cross-country component of a 3-day event and prior to the show jumping test.

Comments on the study

This simply designed study provided further evidence that orally administered glucose does not enhance the rate of muscle glycogen synthesis in horses after glycogen-depleting exercise. However, it is likely that factors other than those associated with glucose delivery via the gastrointestinal tract also constrain post exercise glycogen storage in horses. Indeed, even in the IV treatment, the rate of net glycogen storage was at least 50% lower when compared to a similar treatment regimen in human subjects. One possibility is lower inherent insulin sensitivity to glucose transport in the skeletal muscle of horses when compared to humans and rodents. This hypothesis is supported by the degree of hyperglycemia achieved during the IV treatment; mean plasma glucose concentrations were about 2-fold higher than those observed in human athletes receiving IV glucose at a similar (or even higher) rate of infusion. In humans and rodents, a post exercise enhancement in insulin sensitivity to glucose transport in skeletal muscle facilitates rapid glycogen replenishment when carbohydrate is ingested (or administered IV). Recent studies in our laboratory demonstrated that insulin sensitivity was not affected by prior exercise that resulted in a significant reduction in muscle glycogen content in horses; the lack of change in insulin sensitivity is consistent with the slow rate of muscle glycogen resynthesis observed in equine studies.

What next?

Fundamental work examining glucose metabolism in muscle in response to exercise and insulin-mediated signals is needed to elucidate constraints on muscle glycogen synthesis. Future studies should also determine whether the addition of other substrates (e.g. vegetable oil, amino acids) to post-exercise feedings can enhance glycogen storage. Studies in humans have shown that the addition of protein or amino acids such as leucine to carbohydrate meals is beneficial when only moderate amounts of glucose (<1 g/kg every 2 h) are ingested; perhaps this approach will also be useful in horses?

Raymond J. Geor

Lacombe, V.A., Hinchcliff, K.W., Kohn, C.W., Reed, S.M. and Taylor, L.E., 2006. Effects of dietary glycemic response after exercise on blood concentrations of substrates used indirectly for muscle glycogenesis. Equine Veterinary Journal Supplement 36: 585-589.

Objective

To study the effects of feeding meals of varying glycemic responses on blood concentrations of substrates used for glycogenesis in horses with exercise-induced glycogen depletion. It was hypothesized that horses fed a diet which induced low glycemic response rely more on gluconeogenesis and on lipid utilization for muscle glycogen synthesis than horses fed a diet that induced a high glycemic response.

Background

This study further explores the effects of post exercise diet on the mix of circulating substrates that may directly or indirectly influence the rate of muscle glycogen synthesis. Studies in humans and rodents have clearly demonstrated that supply of glucose is the most important determinant of muscle glycogen replenishment. Is this also true in horses? It has been suggested that low availability of lipid metabolites during the post exercise recovery period may limit muscle glycogenesis by redirecting glucose away from glycogen synthesis to support immediate energy needs through the tricarboxylic acid cycle (Hyyppa *et al.*, 1997), i.e. available glucose is partitioned to oxidation rather than storage. Therefore, diets that directly or indirectly (e.g. via production of volatile fatty acids by microbial fermentation) increase the availability of lipid metabolites could support glycogen synthesis.

Overview of the study

In a 3-way crossover design, 7 fit Standardbred horses received each of 3 isocaloric diets for 72 h after exercise (3 consecutive days, as described above by Jose-Cunilleras *et al.*, 2006) that depleted muscle glycogen by more than 75%. The diets were: (1) a high glycemic response (HGI) diet comprised of 21% mixed hay, 37% cracked corn, 20% cracked barley and 20% oats (with estimated starch of 51% on a dry matter [DM] basis; (2) a low glycemic response (LGI) diet comprised of mixed

timothy-alfalfa hay, with an estimated starch content of 4.3% DM; and
(3) a mixed diet consisting of 66% mixed hay and 33% cracked corn
(26.4% starch), which was considered the control (CON) diet. The DE
of the 3 diets was 0.041 Mcal/kg bwt, which corresponded to the daily
energy requirements of horses in light work. The ration was divided
into 3 equal portions and fed every 8 h for 72 h. During the 3 day
glycogen-depleting exercise protocol horses were fed ~8.5 kg of the
mixed hay and 2 kg of mixed grain, for an estimated DE intake of 19
Mcal/horse/day. Samples of middle gluteal muscle for measurement
of muscle glycogen content were collected before exercise (Day 1),
within 10 min of completion of exercise (Day 3), and then daily during
the treatment period. Blood samples for measurement of plasma
lactate, serum nonesterified fatty acids (NEFA), plasma triglyceride and
glycerol concentrations were collected before and immediately after
exercise, then every 60 min for 12 h after exercise and subsequently
at 24, 48 and 72 h.

Main findings

Feeding the HGI diet resulted in significantly higher rate of muscle
glycogen synthesis when compared to the LGI and CON diets. Plasma
glycerol, triglyceride and lactate and serum NEFA were significantly
higher after compared to before exercise. Overall, there was no
significant effect of diet on these variables over the 72 h treatment
period, although there was a tendency for glycerol, triglyceride and
NEFA concentrations to be lower in the LGI and CON diets during the
first 6 h after exercise. It was concluded that horses fed low glycemic
diets had limited lipid utilization without a substantial shift of substrate
utilization toward gluconeogenesis, factors that could have contributed
to the slower rate of muscle glycogen synthesis compared to horses
fed the HGI diet.

Practical interest

In an earlier publication, this research group reported that the
HGI diet enhanced the rate of muscle glycogenesis in horses after
glycogen-depleting exercise, with almost complete replenishment of
glycogen stores in HGI, but not in LGI or CON, by 72 h of recovery
(Lacombe *et al.*, 2004). The authors suggest that performance horses
undertaking regular heavy exercise could benefit from feeding of high

glycemic meals after exercise to hasten muscle glycogen synthesis. Some equine nutritionists (including this author) might question this advice given the very high starch (~51%), low fiber (only 20% hay on a DM basis) content of the HGI diet. As mentioned, horses have a limited capacity for starch digestion in the small intestine and the feeding of large starch-rich meals (>2.5-3.0 kg for a 450-500 kg horse) may result in the delivery of undigested starch to the hindgut and, potentially, increased risk of intestinal disturbances associated with rapid fermentation of starch in the caecum and large colon. The authors reported no adverse effects of the HGI diet in the horses of the present study, albeit at only a moderate level of daily DE intake. Certainly, such a high starch ration should not be fed to horses not previously adapted to starch-rich diets.

Comments on the study

This study did not shed light on the actual contributions of lipid metabolism to overall energy expenditure, nor the rate of gluconeogenesis in relation to dietary treatment and muscle glycogenesis. Use of stable isotope tracer methodology (which is expensive!) would facilitate these measurements. Analysis of the lipid content in muscle biopsies samples might also yield useful information on the contribution of lipids to energy metabolism during recovery from exercise.

Summary

The three studies summarized here, in combination with earlier studies examining muscle glycogenesis in horses, have shed light on factors limiting the rate of muscle glycogen replenishment in horses after glycogen-depleting exercise. These include:

- A limited capacity for digestion of starch and other forms of glucose in the small intestine, which may limit intestinal glucose uptake and the systemic availability of glucose for glycogen synthesis. This hypothesis is supported by the observation that muscle glycogen storage during the first 24 h after exercise is enhanced when glucose is administered IV but not after oral administration of glucose or ingestion of high-starch meals.
- Constraints on glucose entry into skeletal muscle cells and slower activity of glycogen synthase when compared to other mammalian species. Further studies are needed for better characterization

of these mechanisms in equine skeletal muscle, including the potential for other nutrients (e.g. leucine) to modulate glucose transport and utilization.

The search for nutritional strategies that enhance and optimize muscle glycogen replenishment in horses after exercise will continue. There is evidence that a starch-rich, high glycemic diet can modestly enhance the rate of glycogen storage, but the merit of this approach should be weighed against the potential adverse effects of high starch diets.

References

Davie, A.J., Evans, D.L., Hodgson, D.R. and Rose, R.J., 1994. The effects of an oral glucose polymer on muscle glycogen resynthesis in Standardbred horses. J Nutr 124, 2740S-2741S.

Davie, A.J., Evans, D.L., Hodgson, D.R. and Rose, R.J., 1995. Effects of intravenous dextrose infusion on muscle glycogen storage after intense exercise. Equine Vet J Suppl. 18, 195-198.

Hyyppa, S., Rasanen, L.A. and Poso, A.R., 1997. Resynthesis of glycogen in skeletal muscle from Standardbred trotters after repeated bouts of intense exercise. Am J Vet Res 58, 162-166.

Ivy, J.L., Lee, M.C., Rozinick, J.T. and Reed, M.J., 1988. Muscle glycogen storage after different amounts of carbohydrate ingestion. J Appl Physiol 65, 2018-2023.

Lacombe, V., Hinchcliff, K.W., Geor, R.J. and Baskin, C.R., 2001. Muscle glycogen depletion and subsequent replenishment affect anaerobic capacity of horses. J Appl Physiol 91, 1782-1790.

Lacombe, V.A., Hinchcliff, K.W., Kohn, C.W., Devor, S.T. and Taylor, L.E., 2004. Effects of feeding meals with various soluble-carbohydrate content on muscle glycogen storage after exercise in horses. Am J Vet Res 65, 916-923.

Nout, Y.S., Hinchcliff, K.W., Jose-Cunilleras, E., Dearth, L.R., Sivko, G.S. and DeWille, J.W., 2003. Effect of moderate exercise immediately followed by induced hyperglycemia on gene expression and content of the glucose transporter-4 protein in skeletal muscle of horses. Am J Vet Res 64, 1401-1408.

News on equine sports science (2005-06)

Arno Lindner
Arbeitsgruppe Pferd, Heinrich-Roettgen-Str. 20, 52428 Juelich, Germany

Introduction

Very many interesting articles related to equine sport science were published in 2005 and 2006. All cannot be mentioned. Thus, I have selected some and grouped them under the following four sections:
1. new tools;
2. lactate guided exercise;
3. training and schooling contents;
4. effects of training and conditioning on young horses.

However, first I like to discuss three aspects that I have come across while reading papers on training and conditioning for this article. Their consideration will make the results of studies much more meaningful. Exemplary references are not mentioned, because there are (too) many.

These aspects are:
1. Use of the same exercise test prescription during a study no matter the duration of the study: Researchers that design a longitudinal conditioning study of longer duration (months or even years) may encounter that the standardized exercise test (SET) used at the beginning is not sufficiently stressful to determine the adaptation of the horses to an effective conditioning programme during the whole study. Under such circumstances researchers have to prescribe a more stressful SET. However, instead of prescribing a different SET it is better to plan on prescribing a SET where all horses have to go through the same initial routine every time they are tested and the needed additional stress is achieved by adding intervals or steps of exercise at higher speeds to the SET. This strategy only allows for quantifying the magnitude of adaptation during the whole study and outweighs the pitfall of the additional time that the horses will need to be exercised during SET for their evaluation. This routine avoids also to stress excessively inexperienced or less well

conditioned horses because the initial SET prescription fits them while more experienced and better conditioned horses need to go the higher workloads too but can cope well with them being adequate for their fitness level.

2. Differentiate correctly between the terms exercise and conditioning or training: An amazing number of papers do not reflect correctly in the title and in the keywords whether the matter examined was the effect of an exercise or of a conditioning / training programme. These are two different worlds! Of much lesser importance is to differentiate perfectly between the terms training and conditioning because often this may not be possible. Marlin and Nankervis (2002) write in their book: 'Physical training aims at improving or maintaining maximum performance, delaying onset of fatigue, improving skills, minimizing injuries, and maintaining willingness and enthusiasm for exercise whilst conditioning means improving athletic performance by inducing changes that can be evaluated with objective and scientific methods'. Both terms simply have a lot of overlap.

3. Define completely the exercise examined: There are three parameters that always should be defined in a study on conditioning: (1) frequency (number of exercises in one session and during a defined time period), (2) duration (duration of an exercise session and of the conditioning period), (3) intensity (speed, or gait where speed is not available, slope, weight carried). It amazes me how much time and resources are put into studies without adequate definition of these exercise parameters. For conditioning studies it is not sufficient to state 'light', 'medium' or 'heavy' exercise. These terms are subjective. Every effort has to be put into defining and measuring them. Otherwise comparisons with other conditioning programmes are impossible. Neither can there be clear explanations for effects.

New tools

The most obvious help for sports science is the introduction of the Global Positioning System devices (GPS). Several papers were published on the subject (Hebenbrock *et al.*, 2005; Kingston *et al.*, 2006; Vermeulen and Evans, 2006). The importance of this tool is that now for riders and drivers not so schooled in riding and driving horses at a set pace for performing standardized exercise tests (SET) it has become easy to met prescribed speeds. And only under such circumstances it is possible to diagnose well the performance ability of horses competing in disciplines that demand good cardiovascular and metabolic fitness. However, Vermeulen and Evans (2006) may prove me wrong using a regular fast speed exercise session. They record during such a session the whole variety of speeds from walking through maximal gallop and the concomitant heart rates of a horse so that the heart rate running speed relation can be plotted nicely and any parameter like v_{200} can be derived to compare between and within horses. Because their horses are asked to run for a short period of time maximally too they are able to determine (often) the maximal heart rate (HR_{max}) also. This enables them to derive vHR_{max} as well and they propose to use this parameter for fitness evaluation of Thoroughbreds. More research on the subject will demonstrate whether vHR_{max} is a valid indicator of competitive performance, but it is an interesting perspective for Thoroughbreds and may be applicable for Standardbred racehorses too.

Hebenbrock *et al.* (2006) concluded that the Fidelak Equipilot was reliable for simultaneous recording of distance, speed and heart rate of eventing horses.

The other 2 groups worked with the Equitronic Technologies device on Thoroughbreds.

Kingston *et al.* (2006) measured speeds and heart rate during 4 months of race training. They mention that the device used showed a speed variation of up to 0.6%. Additional interesting information of the study was the observation that the heart rates between horses differed to a given workload and that this difference remained in some horses throughout training.

Vermeulen and Evans (2006) used the GPS with heart rate recorder to monitor Thoroughbreds during typical fast exercise training sessions. They found that vHR_{max}, v_{200} and HR_{max} (velocity at maximal heart rate, velocity at 200 beats /min and maximal heart rate) could be reliably determined repeating the same exercise on consecutive days (12 horses; less than 2% variation). In addition, repeating the measurement on 11 two-yrs old Thoroughbreds after 4 weeks of training demonstrated a significant improvement of vHR_{max} and v_{200}

but not HR_{max}. HR_{max} was on average at 215 beats/min and vHR_{max} at 14.9 m/s irrespective of age group.

The other tool that may in future be of use in the field is a portable oxygen consumption (VO_2) measuring device (Cosmed K4b^2). Since long this has been tried without success. At least in this case the VO_2 consumption measurements compared to the standard measurement procedure were well reproduced on a treadmill having horses exercise until fatigue (Art *et al.*, 2006). Because some results indicate that the mask used for measuring may induce changes in the breathing strategy of the horses the authors concluded that it may not be scientifically valid to compare the data with other systems but that it is okay to do so with data obtained with the same system. This holds especially for the VO_2 measurements.

Lactate guided exercise

I am doing research on this subject since 1993. Several studies have been published on it. I like to comment on an aspect that has to be more acknowledged: different SET prescriptions mean different values of the same parameters measured regardless of these being derived from the blood (plasma) lactate running speed, heart rate running speed or oxygen consumption speed relationships. This is not important as long as the objective of their determination is to diagnose performance. Just stick to one SET prescription, if the repeatability of the values is good. However, when the objective is to determine them for guiding exercise intensity then it is important: the same parameter may mean a very different value!

True is that the effect of changing the duration of steps in SET has a rather small effect on the parameters derived from the heart rate running speed relation compared to the effects on the blood lactate running speed relation. The reason for this is that heart rate does not reflect duration but speed of exercise only whilst the blood lactate concentration is affected by both exercise parameters. This holds for the range of time that we care for when designing SETs.

We quantified the effect of changing the duration of the steps in the SET some years ago (Köster, 1996). Horses were submitted in a randomized order on a treadmill set at 6% slope to SETs with steps of 1, 3 and 5 minutes duration (one duration of steps during one SET only). The initial speed was 6 m/s for all SET and speed was increased by 0.5 m/s until horses had a blood lactate concentration of 4 mmol/l or more. In Figure 1 is shown the example of the blood lactate running speed curves of one horse running the SET with steps of differing duration. In the SET with the steps of 1 min duration the horse ran 11 steps before reaching 4 mmol/l, in the SET with 3 min steps it ran 6, and in the SET with 5 min steps 4 steps. The respective mean v_4 values were 10.5 m/s, 8.2 m/s and 7.5 m/s (Figure 1). Thus, it has to be realized that the information published by us is specific for: (1) SET steps of 5 min duration, and (2) the initial speed of SET and the speed increment from step to step is such that the horses have to run at least 4 steps before reaching or being above 4 mmol lactate

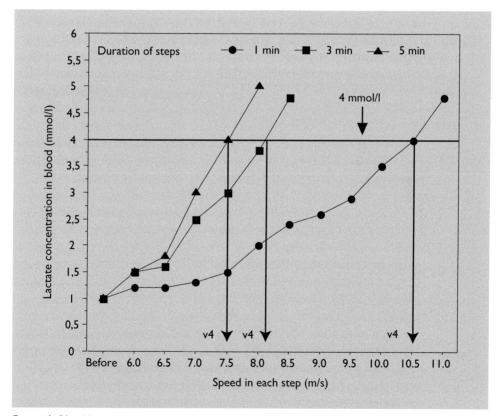

Figure 1. Blood lactate running speed curves of a horse submitted on different days to SETs with steps of 1, 3 and 5 min duration.

per l of blood. When this is not the case we add steps at higher speeds until the blood lactate concentration is above 4 mmol/l. The longer duration of steps is on purpose because we want to apply the research data in the field and often track conditions in the field do not allow for high speeds to be run without incurring a higher risk for injury. Thus, using longer duration steps v_4 can be determined at much lower speed and injury risk is reduced. However, most practitioners not knowing about these details will want to save on time and reduce the duration of steps having then to increase the speed of the steps to achieve the same lactate concentrations as we do. This means that the v_4 or other values of parameters that they derive from the blood lactate running speed relation are higher than ours and therefore can not be used to exercise horses with the same duration and maybe frequency as for the parameters described in our publications.

The same as for the parameters derived from the blood lactate running speed relation holds for those derived from the oxygen consumption running speed relation. This was demonstrated by Rose *et al.* (1990). They examined the effects of different SET prescriptions on VO_{2max} values and observed that the SET prescription influences markedly the VO_{2max} value.

After this introduction into the subject of lactate guided exercise conditioning I like to review shortly the findings of 3 papers on the subject.

Rivero *et al.* (2006 and 2007) published information on different aspects of the same 6 Thoroughbred horses (2-3 yrs old) exercised for periods of 3 weeks every second day on a treadmill for 5, 15 or 25 minutes at their individual $v_{2.5}$ or v_4 (thus a total of 11 exercise sessions during each conditioning period). The uniqueness of this study is that all horses went in a fully randomized order through all the duration and speed combinations of exercise. Between consecutive conditioning periods were 10 to 14 days without gallop exercise. One aspect of the study that led to discussions with reviewers was the possible carry-over effect from one conditioning period to the other/s. This argument cannot be neglected. However, because of the complete randomization of the conditioning programmes the possible carry-over effects should have affected equally all results and therefore it can be expected that it may not have affected them much.

Gluteus medius muscle biopsies of the horses were taken at 2 and 6 cm depth before and after each conditioning period and analyzed for myosin heavy chain content, fibre-type composition, fibre size, capillary numbers, and fibre histochemical oxidative and glycolytic capabilities. Conditioning adaptations were similar in nature, but they varied greatly in magnitude among the different conditioning protocols. Overall, the use of v_4 as the exercise intensity for 25 min elicited the most consistent adaptations in muscle, whereas the minimal conditioning stimulus evoking a significant change was identified with exercises of 15 min at $v_{2.5}$. Within this range, muscular adaptations showed significant trends to be proportional to exercise load of specific conditioning programs. The practical implication of this study is that exercises for 15 to 25 min/d at velocities eliciting blood lactate concentrations between 2.5 and 4 mmol/l can improve in the short-term (3 wks) the muscular stamina in Thoroughbreds, and exercises of 5 to 15 min at v_4 are necessary to enhance muscular features related to strength (hypertrophy).

D´Angelis *et al.* (2005) submitted 12 Purebred Arabians to 90 days of conditioning on a treadmill at 80% of v_4. The v_4 was determined every 4 weeks with a SET consisting of 4 steps of 1 min duration with the treadmill set at 6% incline at 2, 4, 6 and 8 m/s. Half of the horses received 75 g of creatine monohydrate daily. In the first month of conditioning horses run 3 x week for 10 km in a mean time of 50 min (200 m/min), in the second 4 weeks the distance was increased to 15 km run in a mean duration of 60 min (250 m/min), and in the last 4 weeks the distance was increased to 20 km to be run on average at 80 min (250 m/min). during these 4 weeks once a week short bursts of speed interspersed throughout an exercise bout were asked too. This conditioning program caused a small hypertrophy of fibres of type I and IIA while the IIX fibres reduced its size. Fibre type distribution was not affected. Obviously the v_4 of the horses between the 4th and the 13th week of the conditioning period was not improved because the speed of exercise was not increased. Therefore, the conditioning programme used does not seem to be very effective to improve endurance of horses. The supplementation of creatine did not make a difference on any muscular variable.

Training and schooling contents

Under training and schooling contents shall be understood tools and methods used to work horses before, during and after exercise sessions to maintain or improve the performance of horses. This section comprises effects of work on a slope, neck hyperflexion, kaatsu, girth materials, recovery as well as methods to reduce fearfulness.

Work on a slope

Takahashi *et al.* (2006) implanted force probes into the superficial digital flexor tendon (SDFT) once in the right and once in the left forelimb of 7 Thoroughbreds. Horses were worked on a treadmill at 3.5 m/s and 9 m/s and at 0, 3 and 8% inclination as soon as horses walked sound after surgery. This was on average 1 hour after application of a sedation reversal drug. Very often the probes could not bear the strong forces acting on them and therefore not all data could be collected. Out of 14 sets of measurements data was analyzed for 9 x trot and 3 x canter in the trailing limb on a slope of 3 and 8% and only 2 x at the canter in the leading forelimb on a slope of 3%. The results showed that: (1) increasing the incline tended to decrease peak force in the SDFT at the trot and for the trailing limb at the canter; (2) The force in the SDFT was unchanged in the leading forelimb at the canter on 3% incline! The authors concluded that because the same workload imposed on an incline increases exercise intensity without increasing force on SDFT the injury risk may be reduced.

Eto *et al.* (2006) worked Thoroughbreds on a treadmill at 0 and 10% incline. Five horses exercised 4 min at 90% of their VO_{2max} (13.9-14.1 m/s at 0% and 8.4-9.4 m/s at 10%) and other eight 12 min at 60% of their VO_{2max} (7.9-11.2 m/s at 0% and 5.4-6.0 m/s at 10%). Regardless of exercise type the glycogen depletion in the different muscle fibres of the gluteus medius muscle was statistically the same (range 20 to 40%). Hopefully in a future study the authors will exercise the horses at 60% of their VO_{2max} at 0 and 10% slope but during 4 and 12 min to separate out the effects of duration, speed and incline. It would be nice to have the results with the same design (4 and 12 min) for horses exercised at 90% of their VO_{2max}, but it is unlikely that horses are able to run at this speed for 12 min. In conclusion the exercises used recruited equally the muscle fibres of the gluteus medius muscle and therefore exercise on a slope allows to reduce the speed without diminishing metabolic workload. In addition, injury risk should be lowered.

Effects of neck hyperflexion

A lot of debate has raised the schooling method of hyperflexing the neck mainly of horses used for dressage sports ('Rollkur' in German language). The subject is controversial because of the assumption that horses submitted chronically to it will have pain and health problems. In addition it is said that horses so exercised are not able to perform the required dressage lessons in the 'classical way'. The latter is believed to be due to its effects on the back and the gait coordination. However, it is recognized that the method seems to increase the 'expressivity' of horses providing for very good grades in competition. Epidemiological studies on its effects are not available, but recently several studies have been published examining cardiovascular and kinematic effects during an exercise session (Gomez et al., 2006; Sloet van Oldruitenborgh-Oosterbaan et al., 2006; Van Breda, 2006; Weishaupt et al., 2006). Sloet van Oldruitenborgh-Oosterbaan et al. (2006) examined the workload and stress of 8 riding-school horses when being ridden deep and round with a draw rein ('Rollkur') and when being ridden in a natural frame with only light rein contact. Workload measured by heart rate and blood lactate concentration was slightly higher when horses were ridden 'Rollkur'. Differences in packed cell volume, glucose and cortisol concentrations were not found. The authors did not observe signs of uneasiness or stress and meant that horses improved their way of moving and were more responsive to their rider when the horses were ridden 'Rollkur'. Van Breda (2006) also measured parameters of stress in 7 recreational trained horses and 5 International Grand Prix level trained dressage horses. There were no differences between groups in heart rate variability 30 min postfeeding in the morning, but in the 30 min after a morning exercise session the Grand Prix level horses showed an increased parasympathetic dominance indicating that they tended to have less acute stress than do horses used for recreational activities and not used to be 'Rollkur ridden'. Gomez et al. (2006) and Weishaupt et al. (2006) had 7 high level dressage horses walking and trotting on a treadmill with their head and neck in 6 different positions. Gomez et al. (2006) examined the effects on thoracolumbar kinematics of the horses without a rider, Weishaupt et al. (2006)

investigated the vertical ground reaction force and time parameters of each limb of the horses being ridden. The head and neck of the horses were in the following positions: HNP1 = head and neck unrestrained (speed-matched control); HNP2 = neck raised, bridge of the nose in front of the vertical; HNP3 = as HNP2 with bridge of the nose behind the vertical; HNP4 = head and neck lowered, nose behind the vertical; HNP5 = head and neck in extreme high position; HNP6 = head and neck forward and downward. The results of the study with unridden horses showed that there is a significant influence of head/neck position on back kinematics. Elevated head and neck induce extension in the thoracic region and flexion in the lumbar region, besides reducing the sagittal range of motion. Lowered head and neck produce the opposite. A very high position of the head and neck seems to disturb most normal kinematics. When the horses were ridden HNP5 had the biggest impact on limb timing and load distribution and behaved inversely to HNP1 and HNP6. Shortening of forelimb stance duration in HNP5 increased peak forces although the percentage of stride impulse carried by the forelimbs decreased. Thus, an extremely high HNP affects functionality much more than an extremely low neck. The authors concluded that their studies provide quantitative data on the effect of head/neck positions on thoracolumbar motion and limb timing as well as load distribution and may help in discussions on the ethical acceptability of the training methods. However, the results do not allow to predict whether a specific riding technique is beneficial for the horse or if it may increase the risk for injury.

Kaatsu

This is a method that combines low intensity resistance exercise with restriction of venous blood flow from the working muscle. Abe *et al.* (2006) examined whether in horses there were measurable increases of the extensor digitalis communis (EDC) muscle and the superficial digital flexor tendon (SDFT) walking 12 Standardbred mares during 10 min 6 d/wk for 2 wk. Six of the mares had an elastic cuff placed at the most proximal position of the forelegs and inflated to a pressure of 200-230 mmHg throughout walking and for 5 min standing thereafter. After the 2 wk of conditioning EDC thickness increased in the Kaatsu group by 3.5% while the thickness of the SDFT did not change. Nothing was altered in the control group. The authors hypothesize that a longer period of conditioning may have affected also SDFT thickness because

182   *Applied equine nutrition and training*

of the larger time it takes this tissue to adapt. At least for rehabilitative purposes Kaatsu may be an interesting method for horses.

Girth materials

Bowers *et al.* (2005) compared the effect of commercially available girth materials and commonly used girth tensions on athletic performance of 7 racehorses run to fatigue on a treadmill. The horses were exercised at speeds to produce 95% of maximal heart rates on 15 occasions using a randomised block design, and girthed with 5 different girths at 3 nominal tensions of 6, 12 or 18 kg. The girths used were a standard elastic race girth, an 'American' elastic race girth, an elastic race girth twice the normal width, a standard canvas race girth and a canvas race girth at twice the normal width. Tensions were measured at peak inhalation and exhalation, recorded at rest and during exercise. In addition the length-tension relationships of five commercially available girths for training and racing of Thoroughbred racehorses were studied. The results showed that the elastic and the 'American' elastic girths produced significantly longer run to fatigue times when compared to the standard canvas girth. Also girths tensioned at 6 kg and 12 kg produced significantly longer run to fatigue times than when girthed at 18 kg. There were significant differences between the commercially available girth types at each tension, but differences were not significant between girths of the same type. Girths with an elastic component reached their peak for maximum extension at 14.5 kg and thereafter their extension declined. This study shows that the type of girth and the tension at which it is applied affects athletic performance. Lower girth tensions and the use of elastic materials in the girth may optimise performance.

Recovery

Dahl *et al.* (2006) designed a study in which they examined the effect of 4 different work intensities after exercise on heart rate, blood lactate concentration and plasma CK activities among other variables. In their introduction they stated that their Standardbred racehorse trainers in France normally recover horses after exercise at 40-50% of maximal heart rate (HR_{max}). To determine HR_{max} they submitted horses to the following workout: 3 min at 500, 570 and 640 m/min each followed by 500 m at maximal speed. Thereafter 10 of the horses were put back into their boxes, 10 were walked for 10 min (100 m/min) and then put back into their boxes, 9 were trotted at 250 m/min and then placed in their boxes, and finally 8 horses were trotted at 420 m/min and then placed in their boxes. There were no differences of the increase of plasma CK activities between the groups within the 60 min of observation after the exercise. Heart rate was lower already 10 min after finishing the recovery period of 10 min after exercise and remained lower during the 60 min observed for the group that was recovered at the fastest pace. The blood lactate concentration in the group with the fastest recovery pace was lower than in all other groups already in the 5th min of the recovery period and remained the lowest throughout the observation period. The speed of 420 m/min represented about 70% of HR_{max} (about 226 b/min), 250 m/min about 60% and 100 m/min about 48% of HR_{max} respectively. Therefore, the results show rather clearly that the higher recovery speed allowed the horses to recover better from exercising and may provide advantages for managing sport horses exercised intensely. It is noteworthy too, that the active recovery period was 10 min long only. This is well within the possibilities in practice.

Methods to reduce fearfulness

Sport horses need to be calm in the sense that they do not pose for themselves and their human team danger because of their innate disposition to flight and fight. Christensen *et al.* (2006) examined which of three different training methods (habituation, desensitisation and counter-conditioning) was most effective in teaching horses to react calmly in a potentially frightening situation. For this purpose they trained 27 naive 2-year-old Danish Warmblood stallions according to 3 methods: (1) horses (n = 9) were exposed to the full stimulus (a moving, white nylon bag, 1.2 x 0.75 m) in 5 daily training sessions until they met a predefined habituation criterion (habituation); (2) horses (n = 9) were introduced gradually to the stimulus and habituated to each step before the full stimulus was applied (desensitisation); (3) horses (n = 9) were trained to associate the stimulus with a positive reward before being exposed to the full stimulus (counter-conditioning). Each horse received 5 training sessions of 3 min per day. Horses trained with the desensitisation method showed fewer flight responses in total and needed fewer training sessions to learn to react calmly to test stimuli. Variations in heart rate persisted even when behavioural responses had ceased. In addition, all horses on the desensitisation method eventually habituated to the test stimulus whereas some horses on the other methods did not. Thus desensitisation appears to be the most effective training method for horses in frightening situations. Further research needs to be done with experienced sport horses that become less well manageable under certain situations because this very well may be a sign of fear (besides pain).

Effects of conditioning on young horses

Several studies focused on examining whether training of young (immature) horses makes a difference in their future sportive ability and health.

Santamaria *et al.* (2005) examined the effects of early training for jumping by comparing the jumping technique of horses that had received early training with that of horses raised conventionally. 40 Dutch Warmblood horses were analyzed kinematically during free jumping at 6 months of age. Then, they were distributed at random into a control group that was raised conventionally and an experimental group that received 30 months of early training. At 4 years of age, after a period of rest in pasture and a short period of training with a rider, both groups were analyzed kinematically during free jumping. Subsequently, both groups started a 1-year intensive training for jumping, and at 5 years of age, they were again analyzed kinematically during free jumping. In addition, the horses competed in a puissance competition to test maximal performance. There were no differences in jumping technique between experimental and control horses at 6 months of age, but at 4 years, the experimental horses jumped in a more effective manner than the control horses. However, at 5 years of age, these differences were not detectable anymore and the experimental horses did not perform better than the control horses in the puissance competition. It seems that specific training for jumping of horses at an early age is unnecessary because the effects on jumping technique are not permanent and jumping capacity is not improved.

In New Zealand an extensive series of examinations were done on a group of seven 2-yr-old Thoroughbreds trained in a so called 'conventional manner' on turf and sand tracks for 13 weeks compared to a group kept on pasture (Firth *et al.*, 2005; Firth and Rogers, 2005a;

Rogers *et al.*, 2005). The horses were exercised 6 days a week once a day. In the first 4 weeks horses cantered slowly (7.5 m/s), the second 4 weeks they cantered fast (8.9 m/s), and in the last 5 weeks of the training period they cantered on average at 8.4 m/s and 2 times a week in addition they galloped (14.6 m/s). The main findings are summarized in the article of Firth and Rogers (2005b). Firth *et al.* (2005) scanned at several sites metacarpal and third metatarsal bones. The bone mineral density of the epiphysis was markedly higher and of the diaphysis was slightly higher in trained compared with untrained horses, but greater bone size in the trained horses had the greatest effect on an index of bone strength. They also observed that the clinical examination and ancillary diagnostic aids currently in veterinary clinical use could not detect early abnormalities in metacarpophalangeal joint cartilage found in both trained and untrained horses.

Rogers *et al.* (2005) describe the effect of the conventional race training on kinematic parameters of the trot. For this purpose the horses were trotted in-hand, 5, 9 and 13 weeks after the beginning of training. None of the linear or temporal parameters measured varied significantly between the observation periods. When data from sound horses were pooled, the training group trotted at a higher mean velocity (4.2 vs. 3.2 m/s) and with a longer stride length (2.8 vs. 2.2 m) than the untrained group. The stride duration was longer (669 vs. 662 ms), stance period was shorter (34.1 vs. 39.3%) and mid-stance was achieved earlier in the stride (12.1 vs. 13.1%) in the trained than the untrained group. A longer swing phase in the trained group was associated with an increase in retraction time (9.2 vs. 7.6%). These data indicate that early race training in young Thoroughbreds was associated with quantifiable changes in linear and temporal kinematic parameters of the trot, which were related to the racing training objective of improving the horse's ability to work at higher velocities.

Edwards *et al.* (2005) determined whether specific treadmill exercise regimens would accelerate age-related changes in collagen fibril diameter distributions in the common digital extensor tendon (CDET) of the forelimbs of young Thoroughbreds. For this purpose 24 female Thoroughbreds were trained for 18 weeks (6 horses; short term) or 18 months (5 horses; long term) on a high-speed treadmill; 2 age-matched control groups (6 horses/group) performed walking exercise only. Horses were 24 ± 1 months and 39 ± 1 months old at termination of the short-term and long-term regimens, respectively and euthanized at this ages to obtain specimens of their mid metacarpal CDET. Collagen fibril mass-average diameter for the older horses was significantly less than that for the younger horses. Exercise did not significantly affect fibril diameter or distributions in either age group, and collagen fibril index did not differ significantly between groups.

References

Abe, T., Kearns, C.F., Manso Filho, H.C., Sato, Y. and McKeever, K.H., 2006. Muscle, tendon and somatotropin responses to the restriction of muscle blood low induced by Kaatsu-walk trining. Eq vet J Suppl 36: 345-348.

Art, T., Duvivier, D.H., van Erck, E., de Moffarts, B., Votion, D., Bedoret, D., Lejeune, J.P., Lekeux, P. and Serteyn, D., 2006. Validation of a portable equine metabolic measurement system. Eq vet J Suppl 36: 557-561.

Bowers, J. and Slocombe, R.F., 2005. Comparison of girth materials, girth tensions and their effects on performance in racehorses. Aust Vet J 83: 68-74.

Christensen, J.W., Rundgren, M. and Olsson, K., 2006. Training methods for horses: habituation to a frightening stimulus. Equine Vet J 38: 439-43.

D'Angelis, F.H., Ferraz, G., Boleli, I.C., Lacerda-Neto, J.C. and Queiroz-Neto, A., 2005. Aerobic training, but not creatine supplementation, alters the gluteus medius muscle. J Anim Sci 83: 579-585.

Dahl, S., Cotrel, C. and Leleu, C., 2006. Optimal active recovery intensity in Standardbreds after submaximal work. Eq vet J Supple 36: 102-105.

Edwards, L.J., Goodship, A.E., Birch, H.L. and Patterson-Kane, J.C., 2005. Effect of exercise on age-related changes in collagen fibril diameter distributions in the common digital extensor tendons of young horses. Am J Vet Res 66: 564-568.

Eto, D., Yamano, S., Hiraga, A. and Miyata, H., 2006. Recruitment pattern of muscle fibre type during flat and sloped treadmill running in Thoroughbred horses. Eq vet J Suppl 36: 349-353.

Firth, E.C. and Rogers, C.W., 2005a. Musculoskeletal responses of 2-year-old Thoroughbred horses to early training. 7. Bone and articular cartilage response in the carpus. N Z Vet J 53: 113-122.

Firth, E.C. and Rogers, C.W., 2005b. Musculoskeletal responses of 2-year-old Thoroughbred horses to early training. Conclusions. N Z Vet J 53: 377-383.

Firth, E.C., Rogers, C.W., Doube, M. and Jopson, N.B., 2005. Musculoskeletal responses of 2-year-old Thoroughbred horses to early training. 6. Bone parameters in the third metacarpal and third metatarsal bones. N Z Vet J 53: 101-112 and erratum in 215.

Gómez Álvarez, C.B., Rhodin, M., Bobbert, M.F., Meyer, H., Weishaupt, M.A., Johnston, C. and van Weeren, R., 2006. The effect of head and neck position on the thoracolumbar kinematics in the unridden horse. Equine vet J Suppl 36: 441-451.

Hebenbrock, M., Due, M., Holzhausen, H., Sass, A., Stadler, P. and Ellendorff, F., 2005. A new tool to monitor training and performance of sport horses using global positioning system (GPS) with integrated GSM capabilities. Dtsch Tierärztl Wochens 112: 262-265.

Kingston, J.K., Soppet, G.M., Rogers, C.W. and Firth, E.C., 2006. Use of global positioning and heart rate monitoring system to assess training load in a group of Thoroughbred racehorses. Eq vet J Suppl 36: 106-109.

Köster, A., 1996. Reproduzierbarkeit von in Belastungstests ermittelten Kennwerten (v_2, v_3, v_4, v_{12}, v_{150}, v_{180}, v_{200}) und deren Beeinflußbarkeit durch die Stufendauer bzw. Streckenlänge bei Pferden auf dem Laufband. Doctoral thesis for veterinary medicine, University of Giessen, Germany.

Marlin, D. and Nankervis, K., 2002. Equine exercise physiology. UK: Blackwell Science Ltd.

Rivero, J.L., Ruz, A., Martí-Korff, S. and Lindner, A., 2006. Contribution of exercise intensity and duration to training-linked myosin transitions in Thoroughbreds. Eq vet J Suppl 36: 311-315.

Rivero, J.L., Ruz, A., Marti-Korff, S., Estepa, J.C., Aguilera-Tejero, E., Werkman, J., Sobotta, M. and Lindner, A., 2007. Effects of intensity and duration of exercise on muscular responses to training of Thoroughbred racehorses. J Appl Physiol. 2007 Jan 25; [Epub ahead of print].

Rogers, C.W., Firth, E.C. and Anderson, B.H., 2005. Musculoskeletal responses of 2-year-old Thoroughbred horses to early training. 5. Kinematic effects. N Z Vet J 53: 95-100.

Rose, R.J., Hodgson, D.R., Bayly, W.M. and Gollnick, P.D., 1990. Kinetics of VO_2 and VCO_2 in the horse and comparison of five methods for determination of maximum oxygen content. Eq vet J 9: 39-42.

Santamaria, S., Bobbert, M.F., Back, W., Barneveld, A. and van Weeren, P.R., 2005. Effect of early training on the jumping technique of horses. Am J Vet Res 66: 418-424.

Sloet van Oldruitenborgh-Oosterbaan, M.M., Blok, M.B., Begeman, L., Kamphuis, M.C., Lameris, M.C., Spierenburg, A.J. and Lashley, M.J., 2006. Workload and stress in horses: comparison in horses ridden deep and round ('rollkur') with a draw rein and horses ridden in a natural frame with only light rein contact. Tijdschr Diergeneeskd 131: 152-157.

Takahashi, T., Kasashima, Y., Eto, D., Mukai, K. and Hiraga, A., 2006. Effect of uphill exercise on equine superficial digital flexor tendon forces at trot and canter. Eq vet J Suppl 36: 435-439.

Van Breda, E., 2006. A nonnatural head-neck position (Rollkur) during training results in less acute stress in elite, trained, dressage horses. J Appl Anim Welf Sci 9: 59-64.

Vermeulen, A.D. and Evans, D.L., 2006. Measurements of fitness in Thoroughbred racehorses using field studies of heart rate and velocity with a global positioning system. Eq vet J Suppl 36: 113-117.

Weishaupt, M.A., Wiestner, T., von Peinen, K., Waldern, N., Roepstorff, L., van Weeren, R., Meyer, H. and Johnston, C., 2006. Effect of head and neck position on vertical ground reaction forces and interlimb coordination in the dressage horse ridden at walk and trot on a treadmill. Equine vet J Suppl 36: 387-392.

Expanded abstracts

The influence of low versus high fibre haylage diets in combination with training or pasture rest on equine gastric ulceration syndrome (EGUS)

Andrea D. Ellis[1], Maarten Boswinkel[2] and Marianne M. Sloet van Oldruitenborgh-Oosterbaan[2]
[1]*School of Animal Rural and Environmental Sciences, Nottingham Trent University, United Kingdom*
[2]*Department of Equine Sciences, Faculty of Veterinary Medicine, Utrecht University, Utrecht, the Netherlands*

The full paper was published in Pferdeheilkunde: Boswinkel *et al.* (2007), see references.

Take home message

A semi-controlled study showed that feeding a high concentrate and low fibre diet will lead to significant coprophagy and bedding eating in horses exercised to medium level. These behavioural changes may have prevented development of EGUS in this treatment group. In addition results highlight the need for further research into the effect of pre-fermented forages (silage/haylage) on development of gastric ulcers.

Introduction

Equine gastric ulceration syndrome (EGUS) is a prevalent disorder in performance horses and foals (Bertone, 2000; Dörges *et al.*, 1995; McClure *et al.*, 1999; Murray *et al.*, 1996; Rabuffo *et al.*, 2002; Vatistas *et al.*, 1999).

Factors that damage the gastric mucosa such as decreased pH due to diet or intensive exercise, and decreased mucosal protection (mucus and bicarbonate reduction) have been implicated as causes for EGUS

(Cambell-Thompson and Merritt, 1990; Murray, 1994). A prevalence of up to 70% of gastric ulceration has been reported in racehorses (Hammond *et al.*, 1986). In addition to intensive workloads, these horses are commonly fed high-concentrate low-roughage diets, which have been implicated as one major factor in the cause of EGUS (Hammond *et al.*, 1986). The effect of different diets on the development of EGUS in non-racing horses has not been investigated previously. The aim of this field study was to test the hypothesis that a low-fibre diet induces a higher gastric ulceration score than a high-fibre diet.

Materials and methods

Gastroscopy was performed to examine the influence of a low fibre (LF) and a high fibre (HF) diet on the presence of gastric ulceration in thirty 3-year old Dutch Warmblood horses following a three months training period and after a three months pasture period. The study was part of a larger project to measure the effect of diet composition and training on digestibility and behaviour.

Horses were blocked according to sex and then randomly assigned into a high fibre (HF) (dry matter (DM) ratio – concentrate: haylage = 1:4, n = 15) and a low fibre (LF (DM ratio – concentrate: haylage = 4:1, n = 14) feed group during the Training Period.

In the first part of the study all horses were stabled individually and fed either an iso-energetic concentrate-forage HF (75% haylage) or LF (25% haylage) diet for sixteen weeks. The concentrate pellets were designed for performance horses and had a starch and sugar content of 32%. The dry matter content of the haylage was 54%. The word 'fibre' within the trial refers to 'structural fibre' which has not been chopped or ground down through processing. Horses were exercised daily throughout this period, after which the first gastroscopy was performed. Exercise was increased from light to medium level over the three months. During the last four weeks horses were exercised daily on a training mill for 72 minutes (21 min. walk, 35 min. trot, 15 min. canter) and fed 1.5 times the energy requirements for maintenance (Net Energy, CVB, 1996).

The second gastroscopy for all horses was performed following a fourteen week Pasture period during which grass was supplemented with haylage due to extremely low summer pasture.

The entire squamous and glandular mucosa of the stomach was examined on each gastroscopy and scored for amount of lesions and severity of lesions according to the method by MacAllister *et al.* (1997). Additionally, a score on the amount of hyperkeratosis of the squamous mucosa was performed ('hyperkeratosis score') and a 'total clinical score' was given.

Results

No severe lesions were present in any of the horses during any of the examinations. Most ulcers were seen in the nonglandular squamous mucosa of the stomach.

After the Training Period, the HF-group showed a significantly higher score for the number of lesions ($p < 0.05$) and for hyperkeratosis ($p < 0.05$) compared to the LF-group. Following the Pasture Period, there was no difference in scores between horses previously assigned to the LF- and HF-groups. Comparing the results of the HF-group after the Training period with the results after the Pasture period, there were no differences. However, the LF-group showed significantly higher scores for 'overall score' ($p < 0.05$), for 'number of lesions' ($p < 0.01$), for 'severity of lesions' ($p < 0.01$) and for 'hyperkeratosis' ($p < 0.01$) after the Pasture period compared with after the Training period.

Additionally, horses of the HF group showed high haylage retention in the stomach after 12 hours fasting, whilst the LF group exhibited marked bedding eating and coprophagia ($p < 0.001$) (Ellis *et al.*, 2006). The results of gastroscopy following the Pasture period with extra haylage feeding for all horses, were similar to the HF horses following the Training period.

Discussion

The gastroscopy study was part of a larger trial. Therefore, several circumstances were sub-optimal to help counter any possible biases within the gastroscopy study. Despite these shortcomings, the number of horses with an identical age and history from birth, made the gastroscopic examinations useful as many previously reported studies on gastric ulcers in sports horses have not had the benefit of identical age, breed, sex distribution, exercise and diet levels for comparison between groups. Moreover, each horse acted as its own control between the two testing periods.

The unexpected results may be related to the retention of the pre-fermented feed (haylage) in the stomach leading to continuing fermentation with increased volatile fatty acid (VFA) production in the stomach. In most gastroscopy studies, a mean period of 10-12 hours fasting, is normally enough for gastric emptying (Vatistas *et al.*, 1999). However, in horses of the HF-group, that were fed a considerable amount of haylage (13 $\pm$ 1 kg Wet Matter per day), a large amount of ingesta was still present in the stomach after a minimum of 12 hours of fasting. This was also contrary to expectation and may have been related to the high haylage content of the diet. An *in vitro* experiment by Nadeau *et al.* (2003) showed a decrease in the stomach barrier function and sodium transport in response to exposure to volatile fatty acids.

The consumption of bedding and coprophagy shown by the LF group may have limited the previously reported high risk of gastric ulceration on a low fibre diet.

References

Andrews, F.M., Bernard, W., Byars, D., Cohen, N., Divers, T., MacAllister, C. and Pipers, F., 1999. Recommendations for the diagnosis and treatment of equine gastric ulcer syndrome (EGUS). Eq. Vet. Educ. 1:122–134.

Bertone, J.J., 2000. Prevalence of gastric ulcers in elite heavy use western performance horses. Proc. Am. Assoc. Pract. 46: 256-259.

Boswinkel, M.A., Ellis, A.D. and Sloet van Oldruitenborgh-Oosterbaan, M.M., 2007. The influence of low versus high fibre haylage diets in combination with training or pasture rest on equine gastric ulceration syndrome (EGUS). Pferdeheilkunde 23: 123-130.

Campbell-Thompson, M.L. and Merritt, A.M., 1990. Basal and pentagastrin-stimulated gastric secretion in young horses. Am. J. Physiol. 259, 1259-1266.

CVB: Centraal Veevoederbureau, 1996. Het definitieve VEP_en VREp_systeem. Dutch Net Energy System for horses; In: Documentatie rapport nr. 15.

Dörges, F., Deegen, E. and Lundberg, J., 1995. Magenläsionen beim Pferd - Hohe Inzidenz bei gastroskopischen Untersuchungen. Pferdeheilkunde 11: 173-184.

Ellis, A.D., Visser, E.K., Schilder, M.B.H. and van Reenen, C.G., 2006. The effect of high fibre versus low fibre diet on behaviour in horses, 40th International Conference of the Society for Applied Ethology, p.42, Cranfield University Press.

Hammond, C.J., Mason, D.K. and Watkins, K.L., 1986. Gastric ulceration in mature Thoroughbred racehorses. Equine Vet. J. 18: 284- 287.

McClure, S.R., Glickman, L.T. and Glickman, N.W., 1999. Prevalence of gastric ulcers in show horses. J. Am. Vet. Med. Assoc. 215: 1130-1133.

MacAllister, C.G., Andrews, F.M., Deegan, E., Ruoff, W. and Olovson, S.G., 1997. A scoring system for gastric ulcers in the horse. Equine Vet. J. 29: 430-433

Murray, M.J., 1994. An equine model of inducing alimentary squamous epithelial ulceration. Dig. Dis. Sci. 39: 2530-2535.

Murray, M.J., Schusser, G.F., Pipers, F.S. and Gross, S.J., 1996. Factors associated with gastric lesions in Thoroughbred racehorses. Equine Vet. J. 28: 368-374.

Nadeau, J.A., Andrews, F.M., Patton, C.S., Argenzio, R.A., Mathew, A.G. and Saxton, A.M., 2003. Effects of hydrochloric, acetic, butyric, and propionic acids on pathogenesis of ulcers in the nonglandular portion of the stomach of horses. Am. J. Vet. Res. 64: 404-412.

Rabuffo, T.S., Orsini, J.A., Sullivan, E., Engiles, J., Norman, T. and Boston, R., 2002. Association between age or sex and prevalence of gastric ulceration in Standardbred reacehorses in training. J. Am. Vet. Med/ Assoc. 221: 1156-1159.

Vatistas, N.J., Snyder, J.R., Carlson, G., Johnson, B., Arthur, R.M., Thurmond, M., Zhou, H. and Lloyd, K.L., 1999. Cross-sectional study of gastric ulcers of the squamous mucosa in Thoroughbred racehorses. Equine Vet. J. 29: 34-39.

An investigation into the efficacy of a commercially available gastric supplement for the treatment and prevention of Equine Gastric Ulcer Syndrome (EGUS)

E. Hatton[1,2], C.E. Hale[2]* and A.J. Hemmings[3]
[1]TRM, Industrial Estate, Co. Kildare, Ireland
[2]Writtle College, Writtle, Chelmsford, Essex, CM1 3RR, United Kingdom
[3]Royal Agricultural College, Cirencester, Gloucestershire, GL7 6JS, United Kingdom
*catherine.hale@writtle.ac.uk

Take home message

Equine Gastric Ulcer Syndrome (EGUS) has been found to be widely prevalent in both racing, and non-racing horses (Murray *et al.*, 1989; Hartmann and Frankeny, 2003; Nieto *et al.*, 2004; Chameroy *et al.*, 2006). Traditional treatments often necessitate drug threrapy (Andrews *et al.*, 1999), which may be expensive, and competition regulations could inhibit their use. GNF™ is a commercially available nutritional supplement, intended for daily feeding to horses with gastric disturbances. This trial investigated the efficacy of the product in the treatment of EGUS and found that horses supplemented with GNF™ for six weeks showed significant ($P < 0.05$) reduction in overall ulcer score. It can therefore be concluded that GNF™ can have effective results at reducing the severity of EGUS in affected horses.

Introduction

With the development of sophisticated methods of gastroscopy in recent years, Equine Gastric Ulcer Syndrome (EGUS) has become widely diagnosed in many animals (Hartmann and Frankeny, 2003; Nieto *et al.*, 2004; Chameroy *et al.*, 2006). It has been suggested that

E. Hatton, C.E. Hale and A.J. Hemmings

prevalence of the condition amongst racing Thoroughbreds may be as high as 90% (Murray *et al.*, 1989), with gastric lesions being identified in 100% of racing animals in some instances (Murray *et al.*, 1996). In non-racing horses, 51% of animals showing signs of gastric disturbance were found to be effected, with a further 37% of apparently healthy horses displaying significant numbers of lesions to be classed as suffering from EGUS (Bullimore *et al.*, 2001). Clinical signs of the disease include weight loss, diarrhoea, decreased appetite, behavioural changes, decreased performance and colic (Murray *et al.*, 1989; Murray *et al.*, 1996; McClure *et al.*, 1999; Vatistas *et al.*, 1999; Bullimore *et al.*, 2001; Nieto *et al.*, 2004).

Purported reasons for the development of EGUS are wide ranging. Bullimore *et al.* (2001) suggest that ulceration arises from imbalances between defensive mechanisms and aggressive factors within the stomach. It is often assumed that excess acid and pepsin secretion in the glandular region of the stomach may be to blame. However, work in human suffers of gastric ulceration have shown many patients to display near normal acid secretion (Grossman *et al.*, 1963). It is therefore highlighted that defensive mechanisms within the stomach are just as vital in the prevention of ulceration. In the glandular region of the stomach, a mucus layer is secreted to protect against autodigestion, bacterial infection etc. (Bullimore *et al.*, 2001). It has been found that bicarbonate ions are secreted into the mucus allowing surface pH to be maintained near neutral, even when luminal pH is below 2 (Quigley and Turnberg, 1987). Mucus also contains the glycoprotein mucin. It is postulated that abnormal variations and molecular characteristics of mucins can compromise permeability of mucus gels, and therefore mucosal defence (Bullimore *et al.*, 2001). Indeed, in human patients, gastric ulceration has been associated with abnormal mucin gene expression and glycosylation (Jass and Roberton, 1994; Filipe and Ramachandra, 1995).

Although it is acknowledged that the majority of EGUS lesions are found to be present in the non-glandular region of the equine stomach, it has recently been noted that a gene homologous to the human $MUC5_{AC}$ is expressed within the equine stomach in both glandular and non-glandular regions (Bullimore *et al.*, 2001). $MUC5_{AC}$ is responsible for the production and expression of neutral mucins, and its presence in both regions of the stomach suggest some mucosal defence in the proximal stomach areas. Therefore, abnormal production of the gene may lead to reduced defence, as highlighted in humans (Bullimore *et al.*, 2001). Other causes of ulceration are thought to be due to excessive acid build up (Jones, 2002), which could lead to unnatural movement of acid in the non-glandular regions. This may be due to mechanical blockage of the stomach. However, this situation may be compounded in meal-fed horses given restricted access to forage, as stomach secretions are continuous, even when the stomach is empty (Frape, 1998). Furthermore, restricted forage intake is also noted to reduce salivary bicarbonate production, and thus reduce the overall buffering capacity within the proximal end of the stomach (Frape, 1998).

Treatment of EGUS is primarily with pharmacological agents that either act as H2 blockers or as acid pump inhibitors (Andrews *et al.*, 1999). Nutritional supplements that have proven their efficacy in scientific trials could potentially be used as adjuncts to veterinary therapy following an initial course of drug treatment, or during periods were conventional drug treatment must be withheld in order to avoid contravening doping regulations. However, it is the prevention of this disease, through management and possible supplementation of gastric treatments, which horse owners could use to ultimately reduce the high prevalence of EGUS.

The aim of this study was to investigate the efficacy of a commercially available gastric supplement, purported to increase mucin production and combine with natural bicarbonate production to produce defensive gels within the stomach.

Materials and methods

Animal management

26 National Hunt Thoroughbred Racehorses were used in the study. All were aged between 4 and 10 years, and all resided on the same yard. All horses were maintained on their original diet, and all horses remained in their normal training routine. Horses were housed in stables, and were each turned out to grass for 1 hour per day. No additional supplements or conflicting medical treatments were given throughout the duration of the trial.

Animal recruitment to the trial

All animals used in the trial were routinely scoped for veterinary purposes. All horses recruited to the trial were to be scoped regardless of the trial, at the request of the trainer. Horses were initially scoped and divided into three categories depending on their ulcer score; Clear (ulcer score 0), n = 5, Mild to Moderate (ulcer score 1-2), n = 14 and Severe ulceration (ulcer score 3-4), n = 7, within which horses were further divided into supplemented (n = 14) and non-supplemented (n = 12) groups. Each group (treatment and control), were blocked to allow equal numbers of each ulcer score to be allocated to either treatment or control. Blocked animals were then assigned to treatments groups randomly. See Table 1.

Table 1. Assignment of horses to treatments. Total numbers in each group were: Clear – Treatment, n = 3, Control, n = 2; Mild to Moderate – Treatment, n = 7, Control, n = 7; Severe – Treatment, n = 4, Control, n = 3.

Horse number	Ulcer score	Treatment (T) or Control (C)	Horse number	Ulcer score	Treatment (T) or Control (C)
1	0	T	14	2	T
2	0	C	15	2	C
3	0	T	16	2	T
4	0	T	17	2	C
5	0	C	18	2	C
6	1	C	19	2	T
7	1	C	20	3	T
8	1	T	21	3	C
9	1	T	22	3	T
10	1	T	23	3	T
11	1	C	24	4	C
12	1	T	25	4	T
13	1	C	26	4	C

Gastroscopy

Gastroscopic examination was performed after a fasting period of 20 hours and water was withheld for 4 hours prior to scoping. The examination was preformed at the request of the owner, by a registered veterinary practitioner. Examination was carried out using a video Med – V.10.300, CCD Cam 3 endoscope with Xenon XL – M180 light source. The scope measures in at 3 meters in length, with an external diameter of 9.8 mm. Each individual horse's gastroscopy was recorded on to a DVD for later consultation. A panel of veterinarians scored each horse and an average score was then given, the scoring system used is outlined in Table 2.

Horses were supplemented for a period of 6 weeks. All horses used in the trial were due to be re-scoped as part of the normal management after this time, therefore the trial was restricted to this length so as not to subject animals to unnecessary endoscoping procedures. Ulceration was then assessed as before. Nine horses were removed from the trial due to unrelated reasons that resulted in the cessation of training, this resulted in overall group numbers being – treatment n = 11 and control n = 6.

Supplementation

The commercially available supplement, GNF™ was administered to treated horses at recommended levels of 100g/day, split between three feeds, for a period of 6 weeks. GNF™ is marketed as a gastric supplement intended for daily feeding to horses prone to gastric disturbances. It

Table 2. Ulcer Scoring System (as proposed by EGUS Council).

Score	Description
0	Epithelium is intact throughout; no hyperemia, no hyperkeratosis- normal
1	Mucosa is intact but there are areas of hyperemia and/or hyperkeratosis (thickening)
2	Small, single or multi-focal erosions or ulcers
3	Large, single or multi-focal ulcers, or extensive erosions/superficial lesions
4	Extensive ulcers, with areas of deep submucosal penetration

is purported that GNF™ will assist in maintaining optimum gut health and function, allowing maximum utilization of feed.

Compositional Analysis of Supplement	Per 100g
Calcium Carbonate	20,000 mg
Magnesium hydroxide	10,000 mg
Seaweed extract (from *Laminaria hyperborea*)	10,000 mg
Fructo-oligosaccharides	10,000 mg
Glutamine	3,800 mg
Threonine	4,720 mg
Excipients and binders (Full fat soya, Kaolin)	41,480 mg

Justification for ingredient inclusion

- Fructo-oligosaccharides (FOS) have recently been classified as prebiotics (Mikkelsen and Jensen, 2004); substances that are recognised to stimulate growth of desirable bacteria (Kapiki *et al.*, 2007), which can result in a positive symbiotic relationship between bacteria and host (Gibson and Roberfroid, 1995). Due to the formation of β-linkages in the monomer chain, FOS are categorised as non-digestible oligosaccharides (NDOs), as such linkages cannot be hydrolysed by endogenous enzymes (Burvall *et al.*, 1979; Oku *et al.*, 1984). As a consequence, FOS can remain available as substrates for microbial populations to utilise (Houdijk *et al.*, 1998), and thus promote increased intestinal efficiency (Mikkelsen and Jensen, 2004).
- Glutamine is implicated in the synthesis of proteins, as a fuel reserve for dividing cells and lymphocytes, (Krebs *et al.*, 1980), and as a mediator in the development of intestinal epithelial cells (Windmueller and Spaeth, 1980; Wu *et al.*, 1995). Glutamine supplementation has proven to increase intestinal performance (Yan and Qiu-Zhou, 2006), and has been found to decrease over-expression of pro-inflammatory genes, thus leading to a reduction in intestinal damage of rats receiving acetic acid supplementation (Fillmann *et al.*, 2007). Glutamine has also been shown to be an essential requirement of extracellular fluid which is involved in the regulation of intentional cell volume changes, which occur as

a result of cellular regulatory pathways (Ernest and Sontheimer, 2007).

- Threonine is an essential amino acid and studies have shown that restriction of this nutrient may limit intestinal mucin synthesis and reduce gut barrier function (Hamard *et al.*, 2007; Faure *et al.*, 2005).

- *Laminaria hyperborea* is a brown kelp or seaweed that has been found to be extremely palatable and provide increased digestible energy sources for sheep kept on the Orkney island of North Ronaldsay (Hansen *et al.*, 2003). *L. hyperborea* contains higher amounts of vitamins, minerals and proteins, when compared to conventional vegetable sources (*ibid.*).

- GNF contains both calcium and magnesium which are recognised as alkaline providers, and have been shown to increase intestinal mucosal integrity (Wang, 2000).

Results

73% of supplemented horses showed an overall decrease in ulcer severity across all categories. This is in comparison to 33% of control horses showing an increase in severity, and a further 33% of control horses eliciting no change in ulcer score.

The data was tested for normality using the Kolomorgorov-Smirnov Test and found to be significant ($P < 0.05$). It was therefore assumed that the data was not normally disturbed, thus a non-parametric test was used. Data was statistically analysed in SPSS using a two tailed Wilcoxon test, which allows unequal group sizes to be analysed. Overall scores (regardless of category) showed a significant reduction over the trial in supplemented horses, compared to control horses ($p < 0.05$) (See Figure 1). Statistical analysis on individual groups was not possible due to resultant low numbers in each, although supplemented horses showed a trend towards reduced ulcer scores. The average ulcer score for the supplemented group decreased from 1.82 to 0.91, whilst the average score for the control group increased from 1.5 to 1.66.

E. Hatton, C.E. Hale and A.J. Hemmings

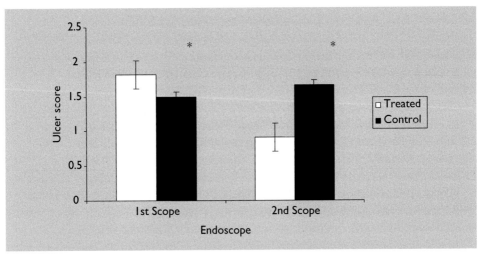

Figure 1. Differences in ulcer score between the first and second gastroscopy exams, in both supplemented and control groups. * Indicates significant differences between the two groups (P<0.05).

Conclusion

This study has proven the effectiveness of GNF™, as a nutritional adjunct in the management of equine gastric ulceration. A significant difference (P < 0.05) in ulceration score was recorded after 6 weeks of supplementation, with the treated group showing significant improvement in comparison to the control group. Due to the exceptionally high prevalence of EGUS in the thoroughbred racehorse sector and the limited availability of clinically proven nutritional feed supplements on the market, the results of this research will provide the industry with an effective nutritional tool in the management of EGUS; in conjunction, were appropriate with proton pump inhibitors such as omeprazole and H2 blockers (e.g. ranitidine, cimetidine).

Acknowledgments

The Authors would like to thank TRM, Ireland, for funding this work and Grey Abbey Veterinary Clinic, County Kildare, Ireland for performing the gastroscopic examinations.

Applied equine nutrition and training

References

Andrews, F.M., Sifferman, R.L., Bernard, W., Hughes, F.E., Holste, J.E., Daurio, C.P., Alva, R. and Cox, J.L., 1999. Efficacy of omeprazole paste in the treatment and prevention of gastric ulcers in horses. Equine Veterinary Journal Suppl., (29), 81-86.

Bullimore, S.R., Corfield, A.P., Hicks, S.J., Goodall, C. and Carrington, S.D., 2001. Surface mucus in the non-glandular region of the equine stomach. Research in Veterinary Science. 70, 149 – 155.

Burvall, A., Asp, N.-G. and Dahlqvist A., 1979. Oligosaccharide formation during hydrolysis of lactose with Saccharomyces lactis lactase (Maxilact®): Part 1. Quantitative aspects. Food Chemistry 4, 243 – 250.

Chameroy, K.A., Nadeau, J.A., Bushmich, S.L., Dinger, J.E., Hoagland, T.A. and Saxton, A.M., 2006. Prevalence of non-glandular gastric ulcers in horses involved in a university riding programme. Journal of Equine Veterinary Science 26, (5), 207 – 211.

Ernest, N.J. and Sontheimer, H., 2007. Extracellular glutamine is a critical modulator for regulatory volume increase in human glioma cells. Brain Research 1144, 231 – 238.

Faure, M., Moennoz, D., Montigon, F., Mettraux, C., Breuille, D. and Ballevre, O., 2005. Dietary threonine restriction specifically reduces intestinal mucin synthesis in rats, Journal of Nutrition 135, (3), 486-491.

Filipe, I.M. and Ramachandra, S., 1995. The histochemistry of intestinal mucins; Changes in disease. In: Gastrointestinal and Oesophageal Pathology. Ed. R Whitehead, Edinburgh. pp. 73 – 95.

Fillmann, H., Kretzmann, N.A., San-Miguel, B., Llesuy, S., Marroni, N., Gonzalez-Gallego, J. and Tunon, M.J., 2007. Glutamine inhibits over-expression of pro-inflammatory genes and down-regulates the nuclear factor kappaB pathway in an experimental model of colitis in the rat. Toxicology. Accepted Manuscript. Available at www.sciencedirect.com. Last accessed 25/04/07.

Frape, D., 1998. Equine Nutrition and Feeding. Blackwell Science.

Gibson, G.R. and Roberfroid, M.B., 1995. Dietary modulation of the human colonic microbiota: introducing the concept of prebiotics. Journal of Nutrition 125, 1401 – 1412.

Grossman, M.I., Kirsner, J.B. and Gillespie, I.E., 1963. Basal and histalog-stimulated gastric secretion in control subjects and in patients with peptic ulcer or gastric cancer. Gastroenterology 45, 14 – 26.

Hamard, A., Mazurais, D., Boudry, G., Le Huerou-Luron, I., Seve, B. and Le Floc'h, N., 2207. Physiological aspects and ileal gene expression profile of early-weaned piglets fed a low threonine diet. Livestock Science 108, 17 – 19.

Hammond, C.J., Mason, D.K. and Watkins, K.L., 1986. Gastric ulceration in mature Thoroughbred horses, Equine Veterinary Journal 18, 284-287.

Hansen, H.R., Hector, B.L. and Feldmann, J., 2003. A qualitative and quantitative evaluation of the seaweed diet of North Ronaldsay sheep. Animal Feed Science and Technology 105, 21 – 28.

Hartmann, A.M. and Frankeny, R.L., 2003. A preliminary investigation into the association between competition and gastric ulcer formation in non-racing performance horses. Journal of Equine Veterinary Science 12, (23), 560 – 561.

Houdijk, J.G.M., Bosch, M.W., Verstegen, M.W.A. and Berenpas, H.J., 1998. Effects of oligosaccharides on the growth performance and faecal characteristics of young growing pigs. Animal Feed Science and Technology 71, 35 – 48.

Jass, J.R. and Robertson, A.M., 1994. Colorectal mucin histochemistry in health and disease: a critical review. Pathology International 44, 487 – 504.

Johnson, B.J., Carlson, G.P., Vatistas, N. and Snyder, J.R., 1994. Investigation of the number and location of gastric ulcerations in horses in race training submitted to the California Racehorse Post mortem Programme, Proceedings of the Annual Meeting of the American Association of Equine Practitioners, 123-124.

Jones, W.E., 2002. Understanding Gastric Ulcers in Horses, Journal of Equine Veterinary Science 7, (22), 330.

Kapiki, A., Costalos, C., Oikonomidou, C., Triantafyllidou, A., Loukatou, E. and Pertrohilou, V., 2007. The effect of a fructo-oligosaccharide supplemented formula on the gut flora of preterm infants. Early Human Development 83, 335 – 339.

Krebs, H.A., Baverel, G. and Lund, P., 1980. Effect of bicarbonate on glutamine metabolism. International Journal of Biochemistry 12, 69 – 73.

McClure, S.R., Glickman, L.T. and Glickman, N.W., 1999. Prevalence of gastric ulcers in show horses. Journal of the American Medical Association 215, 1130 – 1133.

Mikkelsen, L.L. and Jensen, B.B., 2004. Effect of fructo-oligosaccharides and transgalacto-oligosaccharides on microbial populations and microbial activity in the gastrointestinal tract of piglets post-weaning. Animal Feed Science and Technology 117, 107 – 119.

Murray, M.J., Grodinsky, C., Anderson, C.W. and Radue, P.F., 1989. Gastric ulcers in horses: A comparison of endoscopic findings in horses with and without clinical signs. Equine Veterinary Journal, Supplement 7, 68-72.

Murray, M.J. and Eichron, E.S., 1996. Effects of intermittent feed deprivation, intermittent feed deprivation with ranitidine administration, and stall confinement with ad libitum access to hay on gastric ulceration in horses, American Journal of Veterinary Research 57, 1599-1603.

Nieto, J.E., Snyder, J.R., Beldomenico, P., Aleman, M., Kerr, J.W. and Spier, S.J., 2004. Prevalence of gastric ulcers in endurance horses – a preliminary report. The Veterinary Journal 167, 33 – 37.

Oku, T., Tokunaga, T. and Hosoya, N., 1984. Nondigestibility of a new sweetener, 'Neosugar', in the rat. Journal of Nutrition 114, 1574 – 1581.

Quigley, E.M. and Turnberg, L.A., 1987. pH of the microclimate lining human gastric and duodenal mucosa *in vivo*: Studies in control subjects and in duodenal ulcer patients. Gastroenterology 92, 1876 – 1884.

Vatistas, N.J., Snyder, J.R., Carlson, G., Johnson, B., Arthur, R.M., Thurmond, M., Zhou, H. and Lloyd, K.L., 1999. Cross-sectional study of gastric ulcers of the squamous mucosa in thoroughbred racehorses. Equine Veterinary Journal Suppl. 29, 34 – 39.

Wang, X., 2000. How calcium from calcium carbonate and milk benefit peptic ulcer patients. Medical Hypotheses 64 (3), 432 – 433.

Windmueller, H.G. and Spaeth, A.E., 1980. Respiratory fuels and nitrogen metabolism *in vivo* in small intestine fed rats. Quantitative importance of glutamine, glutamate and aspartate. Journal of Biochemistry 225, 107 – 112.

Wu, G., Flynn, N.E., Yan, W. and Barstow, D.G., 1995. Glutamine metabolism in chick enterocytes: absence of pyrroline-5-carboxylase synthase and citrulline synthesis. Biochemistry Journal 306, 717 – 721.

Yan, L. and Qiu-Zhou, X., 2006. Dietary glutamine suuplementation improves structure and function of intestine of juvenile Jian carp (*Cyprinus carpio var. Jian*). Aquaculture 256, 389 – 394.

Changes in plasma metabolites concentrations and LDH isoenzyme activities in Thoroughbred racehorses after racing

Akihiro Mori[1], Hiroyuki Tazaki[1], Nobuko Mori[1], Kieko Tan[1], Yukino Sakamoto[1], Masaru Hosoya[2], Ryuma Nuruki[2] and Toshiro Arai[1]
[1]Department of Veterinary Science, School of Veterinary Medicine, Nippon Veterinary and Life Science University, Tokyo, Japan
[2]YCL Horse Clinic, Saitama, Japan

Take home message

Lactate and pyruvate concentration as well as LDH activity and its isoenzyme pattern changes in plasma of Thoroughbred racehorses after racing. These parameters may be useful to evaluate the energy metabolism in racehorses for racing.

Introduction

The concentration of plasma metabolites in race horses changes after racing. Plasma lactate concentration has been used as an indicator of exercise stress in racehorses (Anderson, 1976). Pyruvate, the end product of glycolysis, is considered to play an important role in regulating some metabolic pathways, including the tricarboxylic acid cycle (TCA cycle), gluconeogenesis and lipogenesis. The reaction between pyruvate and lactate is catalyzed by lactate dehydrogenase (LDH), which is a tetrameric molecule made up of four subunits of the two parent molecule, the H and M subunits, and each tissue has a characteristic composition of LDH isoenzymes. The isoenzyme patterns of LDH are different in plasma and leukocytes of each animal species (Arai *et al.*, 2003) and the isoenzyme patterns are used frequently as a diagnostic marker for metabolic diseases. In the present study, the concentrations of plasma glucose, triglyceride, cholesterol, free fatty acid, pyruvate and lactate as well as the LDH activity and its isoenzyme pattern were measured to investigate the significance of these parameters for evaluating the energy metabolic conditions in thoroughbred racehorses before and after racing.

Material and methods

Six thoroughbred racehorses (3 female and 3 male, 3-7 years old) kept and trained at the Noda Training Center of the Urawa Race Course, National Association of Racing (Saitama, Japan) were examined. Each racehorse was usually exercised for nine of 10 days. The 9 days work included 3 days of fast galloping at 13-17 m/s for 1000-1200 m and 6 days of slow work at 6-8 m/s for 1500-2000 m, including warming up and cooling down for 2 h. All race horses were maintained on grass supplemented with good quality hay and concentrate diet. Blood samples were taken from each horse 20-24 h before and 24-26 h after their races. Blood was withdrawn from the jugular vein into heparinized tubes, with the horses at rest and unfasted, between 12:00 and 14:00 h. The plasma was recovered by centrifugation at 4 °C. Plasma glucose, triglyceride (TG), free fatty acid (FFA), lactate and pyruvate concentrations were measured by previously reported methods (Sako *et al.*, 2007). P/L ratio was calculated as pyruvate concentration divided by lactate concentration. LDH activity was measured with pyruvate (LDH-P) and lactate (LDH-L) as substrate at 25±2 °C. The enzyme unit (U) was defined as 1 μmol of substrate degraded per min. LDH-L/P ratio was calculated as LDH-L activity divided by LDH-P activity. The LDH isoenzyme pattern was analyzed using Epalizer II (Helena Laboratory, Tokyo, Japan). Values are presented as mean (SD) and the differences between means were analyzed by Student's *t*-test. Differences were considered significant when the P value was <0.05.

Results

The plasma glucose, TG and FFA concentrations did not differ before and after racing. Plasma pyruvate concentrations decreased and lactate concentrations significantly increased in the racehorses after racing. The P/L ratio decreased significantly after racing. The elevation of LDH-P activity was larger (42%) than for LDH-L (22%). Thus, the LDH-L/P ratio decreased after racing. In plasma of racehorses, LDH-2, -3 and -4 isoenzymes were dominant and LDH-1 and -5 isoenzymes were minor. Percentages of LDH-4 and -5 after racing were higher than those before racing. Percentages of LDH-2 and -3 isoenzymes decreased after racing.

Discussion and conclusions

After racing, plasma lactate concentrations increased, whereas pyruvate concentrations decreased, as well as the plasma P/L ratio. P/L ratio in herbivorous animals is higher than those in dogs and cats (Sako *et al.*, 2007). These differences in P/L ratios between animal species are considered to be due to differences in diets supplied to animals. As the P/L ratio is considered to be a very sensitive parameter of the metabolic changes in animal tissues, P/L ratio may be useful to evaluate the changes in the physical condition of racehorses during training. Pyruvate and lactate are substrates for LDH and changes in plasma concentrations of these substrates influence the LDH activity. LDH-P activity reflects the total activity of LDH, whereas LDH-L activity is considered to reflect the activities of LDH-1 and -2 isoenzymes, which are inhibited by high concentrations of pyruvate and are dominant in aerobic tissues such as heart muscle. As a high LDH-L/P ratio is observed in tissues in which energy metabolism is accelerated, changes in LDH-L/P ratio may be useful to assess recovering from fatigue in racehorses.

Plasma LDH isoenzyme pattern reflects the origin of LDH leaked from tissues. Usually LDH-2, -3 and -4 isoenzymes are dominant in plasma of horses. LDH-4 and -5 isoenzymes which are dominant in skeletal muscle increased slightly in plasma of racehorses after racing. This elevation in LDH-4 and -5 isoenzymes may reflect slight injury or stress exerted in skeletal muscle during racing.

In conclusion, plasma pyruvate and lactate concentrations and their ratios as well as LDH activities and their isoenzyme patterns changed in Thoroughbred racehorses during racing. These parameters may be useful to evaluate adaptations in energy metabolism and physical condition in racehorses.

References

Anderson, M.G. 1976: The effect of exercise on the lactic dehydrogenase and creatine kinase composition of horse serum. Res. Vet. Sci. 20: 191-196.

Arai, T., Inoue, A., Uematsu, Y., Sako, T. and Kimura, N., 2003. Activities of enzymes in the malate-aspartate shuttle and the isoenzyme pattern of lactate dehydrogenase in plasma and peripheral leukocytes of lactating Holstein cows and riding horses. Res. Vet. Sci. 75. 15-19.

Sako, T., Urabe, S., Kusaba, A., Kimura, N., Yoshimura, I., Tazaki, H., Imai, S., Ono, K. and Arai, T., 2007. Comparison of Plasma Metabolite Concentrations and Lactate Dehydrogenase Activity in Dogs, Cats, Horses, Cattle and Sheep. Veterinary Research Communications 31, No. 4. (May 2007): 413-417.

Changes in enzyme activities in peripheral leukocytes of Thoroughbred racehorses after racing

Toshiro Arai[1], Akihiro Mori[1], Kieko Tan[1], Nobuko Mori[1], Yukino Sakamoto[1], Masaru Hosoya[2] and Ryuma Nuruki[2]
[1]Department of Veterinary Science, School of Veterinary Medicine, Nippon Veterinary and Life Science University, Tokyo, Japan
[2]YCL Horse Clinic, Saitama, Japan

Take home message

Malate dehydrogenase (MDH) activity in the peripheral leukocytes of Thoroughbred racehorses decreased after racing. The peripheral leukocytes cytosolic ratio of MDH/lactate dehydrogenase (M/L ratio) may be a useful indicator to evaluate energy metabolism in racehorses.

Introduction

In racehorses undergoing continuous training, activities of D-glucose transport and glycolytic enzymes in blood cells increased significantly compared with those in riding horses (Arai *et al.*, 1994). Malate dehydrogenase (MDH) activities in the malate-aspartate shuttle are significantly increased in racehorses during exercise training (Arai *et al.*, 2002). Activities of lactate dehydrogenase (LDH) as the cytosolic marker enzyme are relatively stable in various metabolic conditions, and the cytosolic ratio of MDH/LDH activity (M/L ratio) is considered to be a useful indicator of energy metabolism in animal tissues. The M/L ratio significantly decreases in the peripheral leukocytes of dogs with severe type 1 diabetes mellitus (Magori *et al.*, 2005).

In the present study, plasma metabolite concentrations and activities of some enzymes related to energy metabolism in peripheral leukocytes of Thoroughbred racehorses were measured before and after racing to investigate their changes and the possibility of using the M/L ratio as an indicator of energy metabolism.

Materials and methods

Five Thoroughbred racehorses (1 female and 4 male, 3-6 years old) kept and trained at the Noda Training Center of the Urawa Race Course, National Association of Racing (Saitama, Japan) were examined. Each race horse was usually exercised for nine out of 10 days. The training consisted of 3 days of fast galloping at 13-17 m/s for 1000-1200 m and 6 days of slow work at 6-8 m/s for 1500-2000 m, including warming up and cooling down for 2 h. All racehorses were maintained on grass supplemented with good-quality hay and concentrate diet. Blood samples were taken from each horse 20-24 h before and 24-26 h after their races. Blood was withdrawn from the jugular vein into heparinized tubes, with the animals at rest and unfasted, between 12:00 and 14:00 h. The plasma was recovered by centrifugation at 4 °C. Leukocytes were isolated from the buffy coat and washed with cold phosphate buffered saline (PBS) and the cytosolic and mitochondrial fractions were isolated. Activities of LDH as a cytosolic marker enzyme, hexokinase (HK) as a rate limiting enzyme of glycolysis, glucose-6-phosphate dehydrogenase (G6PD) as a rate limiting enzyme of the pentose phosphate pathway and MDH and aspartate aminotransferase (AST) as pivotal enzymes in the malate-aspartate shuttle, were measured. The enzyme unit (U) was defined as 1 μmol of substrate degraded per min. Enzyme activities were measured at 24-26 °C. Plasma glucose, triglyceride (TG) and immunoreactive insulin (IRI) concentrations were measured using commercial kits. Values are presented as means (SD) and the differences between means were analyzed by Student's *t*-test. Differences were considered significant when the P value was < 0.05.

Results

The plasma glucose, TG and IRI concentrations and enzymes activities were not different before and after racing. The cytosolic HK and MDH activities in leukocytes decreased to 70 and 80% of the values before the races, respectively, whereas the average LDH activity in leukocytes after the races were 15% higher than those before racing. There were no differences between cytosolic G6PD and AST activities in leukocytes after racing compared to the activities before races. The M/L ratio in all race horses decreased significantly to 70% of the values before the races. Activities of MDH and AST in the mitochondrial fractions decreased significantly after the races compared to those before the races.

Discussion and conclusions

The five horses competed in different races with different distance and classes and the horse with the highest M/L ratio won its race. The other four horses did not win their races. Cytosolic MDH has various roles in insulin secretion and in growth of hepatocytes. Decreased M/L ratio owing to decreasing MDH activities is considered to reflect the decline of energy metabolism in tissues. The M/L ratio may be an indicator to evaluate the degree of exhaustion in racehorses. Plasma LDH, MDH and AST activities did not differ before and after racing. Plasma lactate concentrations and LDH and creatine phosphokinase activities have been used as indicators of exercise stress in racehorses (Anderson, 1976). These parameters can be a good indicator of exercise stress, but less effective in assessing the metabolic conditions in racehorses. In the present study, cytosolic MDH activities and M/L ratio in leukocytes were considered to be more sensitive indicators to evaluate the physical conditions of racehorses compared to LDH, MDH and AST activities in plasma. MDH and AST activities in mitochondrial fractions of leukocytes of racehorses decreased significantly after racing. A decrease of the mitochondrial MDH activity is considered to reflect less production of ATP in mitochondria. Activity of mitochondrial MDH is regulated by concentrations of malate, oxaloacetate, citrate and NAD (Gelpi *et al.*, 1992). The mechanism of the decrease of mitochondrial MDH activity in leukocytes after racing is not clear. In the present study, activity of HK as a rate limiting enzyme in glycolysis decreased significantly in the leukocytes after racing. Decrease in glucose utilization are a common finding in diabetic dogs with insulin deficiency and racehorses after racing. The relationship between glucose utilization and MDH activity should be studied in more racehorses before and after racing.

In conclusion, MDH activity decreased significantly in peripheral leukocytes in racehorses after racing, and the cytosolic M/L ratio was considered to be a useful indicator of the energy metabolic conditions of racehorses.

References

Anderson, M.G., 1976: The effect of exercise on the lactic dehydrogenase and creatine kinase composition of horse serum. Res. Vet. Sci. 20: 191-196.

Arai, T., Washizu, T., Hamada, S., Sako, T., Takagi, S., Yashiki, K. and Motoyoshi, S., 1994. Glucose transport and glycolytic enzyme activities in erythrocytes of two-year-old thoroughbreds undergoing training exercise. Vet. Res. Commun. 18: 417-422.

Arai, T., Hosoya, M. and Nakamura, M., 2002. Cytosolie ratio of malate dehyrogenase/ lactate dehydrogenase activity in peripheral leukoeytes of race horses with training. Res. Vet. Sci. 72: 241-244.

Gelpi, J.L., Dordal, A., Montserrat, J., Mazo, A. and Cortés, A., 1992. Kinetic studies of the regulation of mitochondrial malate dehydrogenase by citrate. Biochem. J. 283: 289-297.

Magori, E., Nakamura, M., Inoue, A., Tanaka, A., Sasaki, N., Fukuda, H., Mizutani, H., Sako, T., Kimura, N. and Arai, T., 2005. Malate dehydrogenase activities are lower in some types of peripheral leucocytes of dogs and cats with type 1 diabetes mellitus., Res Vet Sci 78: 39-44.

Forage conservation methods - impact on forage composition and the equine hindgut

Cecilia E. Müller and Peter Udén
Swedish University of Agricultural Sciences, Department of Animal Nutrition and Management, Kungsängen Research Centre, SE-753 23 Uppsala, Sweden
Cecilia.Muller@huv.slu.se

Take home message

Grass from the same grass crop conserved as hay, haylage or silage differed in several chemical and microbial variables, but had similar impacts on the microbial and chemical composition in right ventral colon and faeces of fistulated horses fed the forages.

Introduction

The use of wrapped forages such as silage and haylage[2] has replaced hay in equine diets in Sweden to a large extent during recent years (Holmquist and Müller, 2002). The reason for this may be multifactorial, but one factor may be the difficulties involved in storing hay correctly (dry and airy). Moist hay is easily subjected to mould growth (Lacey, 1989). Moulds produce both spores and mycotoxins, which may have detrimental effects on equine health as mould spores play a large role in the aetiology of recurrent airway obstruction (Robinson *et al.*, 1996), and mycotoxins have been shown to seriously impair horses' health (Asquith, 1991). Wrapped forages are less prone to mould provided that the plastic seal is intact (McNamara *et al.*, 2002; O'Brien *et al.*, 2007). Also, Vandenput *et al.* (1997) showed that both haylage and silage contained less respirable particles than well-managed hay. However, the preservation methods of silage, haylage and hay also produce differences in chemical and microbial composition of the forage. Silage

---

[2] In this paper, haylage is defined as silage with a dry matter level exceeding 500 g/kg (Gordon *et al.*, 1961)

generally contains a higher amount of lactic acid, lower concentration of water soluble carbohydrates (WSC) and a lower pH than haylage, as the activity of lactic acid bacteria (LAB) are restricted in higher dry matter (DM) levels (Field and Wilman, 1996). The concentration of WSC is generally highest in hay. Investigations concerning the influence of these differences on the hindgut fermentation in horses have, to the authors knowledge, not been published previously.

Materials and methods

The same crop, a first cut permanent grass ley, was used for production of silage (343 g DM/kg), haylage (548 g DM/kg) and barn-dried hay (815 g DM/kg). Silage and haylage was inoculated with freeze-dried LAB plus sodiumbenzoate and potassiumsorbate (Lactisil Horse Plus, Medipharm, Kågeröd, Sweden), and ensiled in large round bales. After 6 months, wrapped bales were opened and rebaled into smaller square bales, and all forages were transported to ENESAD, Dijon, France.

A feeding experiment with four fistulated horses was performed using a change-over design, so that all horses were fed all forages. Horses were only fed the experimental forages, salt and minerals. The daily forage was divided in two parts, with 0.4 fed at 8.00 in the morning, and 0.6 at 18.00 in the evening. After being fed one of the forages for 21 days, samples from right ventral colon and faeces were taken (four hours after feeding the morning meal). A kinetic study of the colon was also performed (data not shown).

Horse samples were analyzed for DM, pH and microbial and chemical composition. Microbial methods used for analysis of colon content and faeces were described by de Fombelle *et al.* (2003), and chemical composition was analysed with HPLC according to Andersson and Hedlund (1983). pH in colon content and faeces was measured immediately after sampling. Forages were sampled during rebaling and during the feeding experiment, and were analysed for microbial and chemical composition according to methods described by Müller and Udén (2007).

The data was analysed using the GLM procedure of SAS (SAS 9.1, SAS Institute, USA) for evaluation of effects of forage, sampling site, horse and interactions between forage and sampling site. Microbial

CFU-values were transformed to ^{10}log before statistical analysis to become normally distributed. Results were P < 0.05 were regarded as significantly different.

Results

The preserved forages differed in several variables (Table 1). Silage differed from haylage in more variables than haylage differed from hay. VFA, lactic acid, alcohols, ammonia-N and counts of lactic acid bacteria were highest and pH and total WSC lowest in silage. Hay contained the highest counts of enterobacteria and mould, and had the highest concentration of sucrose and fructans.

The differences in the conserved forages did not influence the microbial and chemical composition in right ventral colon or faeces differently (Table 2, P > 0.21), and there were no interactions between forage and sampling site at Day 21. In general, differences between sampling sites (right ventral colon and faeces) were found for DM content, counts of lactobacilli and lactate utilizing bacteria which were higher in faeces, and for pH, acetic acid, propionic acid, butyric acid and total organic acids which were higher in colon samples (Table 2). Individual horses differed slightly in a few variables, but these differences were very small and were not considered to have had any major influence on the results (data not shown).

Discussion

The differences found in composition between hay, haylage and silage were typical for the conservation methods employed (Gordon *et al.,* 1961; Müller and Udén, 2007).

Changes in the equine hindgut such as decreases in pH and acetate and increases in propionate, butyrate and lactate have been previously reported in connection with increased amounts of or inclusion of concentrates in the diet (e.g. Hintz *et al.,* 1971; Kern *et al.,* 1973; Julliand *et al.,* 2001), onset of laminitis (e.g. Rowe *et al.,* 1994) or treatment with antibiotics (Kropp, 1991). Such changes in the colon or faeces did not take place when any of the forages in this study were fed.

Table 1. Chemical composition and microbial counts in preserved forages (n=9, except for microbial analysis where n= 4).

Variable	Silage	Haylage	Hay	SEM	P
Chemical composition					
Dry matter, g/kg	343[a]	548[b]	815[c]	11.1	<0.0001
Crude protein, g/kg DM	176[a]	151[b]	165[a]	4.6	0.0004
Total water soluble carbohydrates, g/kg DM	80[a]	126[b]	116[b]	7.0	<0.0001
Glucose, g/kg DM	31[a]	49[b]	42[b]	3.1	0.0005
Fructose, g/kg DM	36[a]	61[b]	42[a]	4.4	0.0007
Sucrose, g/kg DM	6[a]	7[a]	15[b]	1.2	<0.0001
Fructans, g/kg DM	2[a]	4[a]	10[b]	1.2	0.0003
Neutral detergent fibre, g/kg DM	421[a]	468[b]	486[b]	6.8	<0.0001
In vitro digestible organic matter, g/kg DM	905	893	895	5.2	0.19
Lactic acid, g/kg DM	43.0[a]	1.3[b]	0.3[b]	2.97	<0.0001
Acetic acid, g/kg DM	4.8[a]	0.8[b]	0.1[b]	0.30	<0.0001
Succinic acid, g/kg DM	2.8[a]	0.6[b]	<0.1[b]	0.20	<0.0001
Total organic acids, g/kg DM	50.6[a]	2.6[b]	0.5[b]	3.36	<0.0001
Ethanol, g/kg DM	9.7[a]	5.9[b]	<0.1[c]	0.63	<0.0001
2,3-butanediol, g/kg DM	4.5[a]	0.4[b]	<0.1[b]	0.28	<0.0001
pH	4.44[a]	5.60[b]	6.02[c]	0.081	<0.0001
Ammonia-N/total N	7.3[a]	2.4[b]	1.5[b]	0.59	<0.0001
Microbial counts[A]					
Lactic acid bacteria	6.6[a]	4.3[b]	<1.7	0.61	0.0007
Enterobacteria	<1.7	<1.7	2.9	0.16	<0.0001
Clostridial spores	<1.7	<1.7	<1.7	-	-
Yeast	2.0	2.3	<1.7	0.63	0.50
Mould	1.7[a]	<1.7	2.6[b]	0.22	0.0008

[a, b, c] Different superscripts within rows indicate significant difference at the P-level listed.
[A] ^{10}log CFU /g FM.

Samples from colon and faeces were not identical, but molar proportions between acetate:propionate:butyrate:valrate were similar for both sample types. Thus, faecal samples may indicate proportions of VFA in right ventral colon, at least in healthy horses fed only forage. However, more studies on the correlation between colon and faecal composition in equines are needed.

Table 2. Chemical and microbial composition of samples from right ventral colon and faeces) in horses fed forages preserved differently (silage, haylage and hay) (Day 21, n=4).

Variable	Silage Colon	Silage Faeces	Haylage Colon	Haylage Faeces	Hay Colon	Hay Faeces	Site SEM	P
Chemical composition								
Dry matter, g/kg	81[a]	233[b]	117[a]	242[b]	50[a]	234[b]	23.6	<0.0001
pH	6.81[a]	6.23[b]	6.64[a]	6.36[b]	6.75[a]	6.07[b]	0.0059	<0.0001
Lactic acid, mM	0.17	0.08	0.08	0.09	0.05	0.11	0.059	0.92
Acetic acid, mM	43.0[a]	23.1[b]	37.2[a]	27.4[b]	47.9[a]	32.8[b]	5.63	0.003
Propionic acid, mM	11.0[a]	8.0[b]	11.8[a]	8.5[b]	12.2[a]	9.9[b]	1.69	0.04
i-Butyric acid, mM	0.8	0.6	0.6	0.9	0.8	0.7	0.26	0.87
Butyric acid, mM	3.6[a]	2.2[b]	3.7[a]	2.5[b]	3.9[a]	3.0[b]	0.57	0.01
i-Valeric acid, mM	0.8	0.6	0.5	0.8	0.8	0.6	0.27	0.89
n-Valeric acid, mM	0.4	0.3	0.3	0.5	0.4	0.4	0.10	0.55
Total organic acids, mM	59.9[a]	34.8[b]	54.2[a]	40.7[b]	66.0[a]	47.9[b]	7.95	0.006
Microbial composition[A]								
Total anaerobic bacteria	7.9	7.6	7.6	8.2	7.7	7.7	0.21	0.71
Cellulolytic bacteria, MPN/ml[B]	5.1	5.2	4.7	4.7	5.1	4.6	0.39	0.74
Lactate utilizing bacteria	6.9[a]	7.4[b]	7.0[a]	7.4[b]	7.0[a]	7.7[b]	0.13	0.003
Lactobacilli	6.2[a]	7.0[b]	5.7[a]	6.4[b]	5.6[a]	6.8[b]	0.09	<0.0001
Streptococci	6.0	5.4	5.2	5.8	5.5	6.5	0.19	0.09

[a,b] Different superscripts within rows indicate difference at the P-level listed.
[A] ^{10}log CFU /g FM.
[B] MPN, most probable number (McGrady).

Cecilia E. Müller and Peter Udén

Acknowledgement

This experiment was financed by The Swedish Farmers' Foundation for Agricultural Research – the Horse Research Committee.

References

Andersson, R. and Hedlund, B., 1983. HPLC analysis of organic acids in lactic acid fermented vegetables. Z. Lebensm. –Unters. Forsch. 176, 440-443.

Asquith, R.L., 1991. Mycotoxicoses in horses. In: Mycotoxins and animal foods. Eds. J.E. Smith and R.S. Henderson. CRC Press Inc., Boca Raton, USA. pp. 679-688.

Field, M. and Wilman, D., 1996. pH in relation to dry matter content in clamped and baled grass silages harvested in England and Wales. Proceedings of the XIth International Silage Conference, Aberystwyth, Wales, UK, 1996, pp. 126-127. Aberystwyth, U.K.: Institute of Grassland and Environmental Research.

de Fombelle, A., Varloud, M., Goachet, A-G., Jacotot, E., Philippeau, C., Drogoul, C. and Julliand, V., 2003. Characterization of the microbial and biochemical profile of the different segments of the digestive tract in horses given two distinct diets. Animal Science 77, 293-304.

Gordon, C.H., Derbyshire, J.C., Wiseman, H.G., Kane, E.A. and Melin, C.G., 1961. Preservation and feeding value of alfalfa stored as hay, haylage and direct-cut silage. Journal of Dairy Science 44, 1299-1311.

Hintz, H.F., Argenzio, R.A. and Schryver, H.F., 1971. Digestion coefficients, blood glucose levels and molar percentage of volatile fatty acids in intestinal fluid of ponies fed varying forage-grain ratios. Journal of Animal Science 33, 992-995.

Holmquist, S. and Müller, C.E., 2002. Problems related to feeding forages to horses. In: Gechie, L.M., Thomas, C. (Eds.) Conference Proceedings XIIIth International Silage Conference, Auchincruive, Scotland, UK, pp. 152-153.

Julliand, V., de Fombelle, A., Drogoul, C. and Jacotot, E., 2001. Feeding and microbial disorders in horses: part 3 – Effects of three hay:grain ratios on microbial profile and activities. Journal of Equine Veterinary Science 21 (11), 543-546.

Kern, D.L., Slyter, L.L., Weaver, J.M., Leffel, E.C. and Samuelson, G., 1973. Pony cecum vs. steer rumen: the effect of oats and hay on the microbial ecosystem. Journal of Animal Science 37, 463-469.

Kropp, S., 1991. Bakteriologische Untersuchungen zur Zusammensetzung der Darmflora des Pferdes und deren Beeinflussung durch Chemotherapeutika (In German). Dissertation. Tierärztlichen Hochschule Hannover. pp. 11-27.

Lacey, J., 1989. Pre- and post-harvest ecology of fungi causing spoilage of foods and other stored products. Journal of Applied Bacteriology 67 (Suppl.), 11S-25S.

McNamara, K., O'Kiely, P., Whelan, J., Forristal, P.D. and Lenehan, J.J., 2002. Simulated bird damage to the plastic stretch-film surrounding baled silage and its effects on conservation characteristics. Irish Journal of Agricultural and Food Research 41, 29-41.

Müller, C.E. and Udén, P., 2007. Preference of horses for grass conserved as hay, haylage or silage. Animal Feed Science and Technology 132, 66-78.

O'Brien, M., O'Kiely, P., Forristal, P.D. and Fuller, H.T., 2007. Quantification and identification of fungal propagules in well-managed baled grass silage and in normal on-farm produced bales. Animal Feed Science and Technology 132, 283-297.

Robinson, N.E., Derksen, F.J., Olszewski, M.A. and Buechner-Maxwell, V.A., 1996. The pathogenesis of chronic obstructive pulmonary disease of horses. British Veterinary Journal 152, 283-306.

Vandenput, S., Istasse, L., Nicks, B. and Lekeux, P., 1997. Airborne dust and aeroallergen concentrations in different sources of feed and bedding for horses. Veterinary Quarterly 19, 154-158.

The application of mathematical models to assess the effect of enzyme treatment or sugar beet pulp on the rate of passage of ensiled lucerne in equids

Jo-Anne MD Murray[1,2,]*, Ruth Sanderson[1], Annette Longland[1], Meriel Moore-Colyer[2], Peter M. Hastie[2] and Catherine Dunnett[3]
[1]IGER, Plas Gogerddan, Aberystwyth, SY23 3EB, United Kingdom
[2]IRS, University of Wales, Aberystwyth, Llanbadarn Fawr, Aberystwyth, SY23 3AL, United Kingdom
[3]Dengie Crops Ltd, Asheldam, Southminster, Essex, CM0 7JF, United Kingdom
*Present address: Royal (Dick) School of Veterinary Studies, University of Edinburgh, Easter Bush, Midlothian, EH25 9RG, United Kingdom

Take home message

It appears that time-dependent models have a greater ability to describe the pattern of faecal marker excretion in ponies compared to time-independent models; however, it was not possible to draw any definitive conclusions from the compartmentalisation data in this study.

Introduction

Enzyme treatment of lucerne silage has been seen to reduce fibre digestibility when fed to ponies (Murray *et al.*, 2007a). Conversely, substitution of lucerne silage with sugar beet pulp (SB) has been seen to enhance the nutritive value of this forage by enhancing total diet digestibility and improving the degradability of the fibrous fraction of the lucerne (Murray *et al.*, 2007b). Since the manipulation of dietary components can affect the site of nutrient digestion and/or digesta passage rate (RoP) through the gastrointestinal tract, both of which have important implications for nutrient availability to the animal, it is conceivable that this may explain the aforementioned findings. Mathematical models developed to describe ruminant faecal excretion data have been employed as a non-invasive method of describing RoP and MRT in the individual gut compartments of horses (Moore-Colyer *et al.*, 2003; McLean, 2001), although studies are limited and far from established. Consequently, the aims of this experiment were to (1) evaluate the ability of the six compartmental models to describe the pattern of faecal excretion in ponies, and (2) select a 'best fit' model for compartmental analysis of the MRT of enzyme-treated or SB-substituted lucerne silage in the different segments of the equid gastrointestinal tract.

Materials and methods

A 3 x 3 Latin-square design experiment was used to evaluate the RoP of three diets; a lucerne silage control (WS), enzyme-treated lucerne (WE3) and lucerne substituted with 300 g kg^1 DM of SB (SB3). The control diet (WS), WE3 diet, and the basal silage in the SB3 diet were labelled with ytterbium chloride (Yb) (Teeter *et al.*, 1984). Labelled feeds (60 g) were offered as an oral (pulse) dose on the first day of each recording period prior to the morning meal and faecal sampling initiated immediately. Yb concentration of feed and faecal samples was determined according to Siddons *et al.* (1985). Faecal excretion data were fitted to two time-independent models; Grovum and Williams (1973) and Dhanoa *et al.* (1985), and four time-dependent models of Pond *et al.* (1988). Corrected R^2 was calculated for all models to evaluate the 'goodness of fit': the model exhibiting the highest R^2 was considered to best describe the data. R^2 values were analysed by two-way ANOVA. Compartmental parameters were analysed by Latin square ANOVA. All statistical analyses were carried out using Genstat 5 (Lawes Agricultural Trust, 1993).

Results

The models of Grovum and Williams (1973), Dhanoa *et al.* (1985) and the G2G1 model of Pond *et al.* (1988) failed to converge with 12 out of the 15 faecal excretion data collected; thus were rejected from further analysis. The R^2 values for the remaining compartmental models are shown in Table 1. The G3G1 model yielded the highest R^2 values and was chosen to compare differences between the marked feeds (Table 2).

Table 1. Accuracy of fit (R^2) of the Pond et al. (1988) G1G1, G3G1 and G4G1 models fitted to faecal excretion data obtained from Yb-marked lucerne silage (WS), enzyme-treated (WE3) and sugar beet pulp substituted (SB3) lucerne silage.

Diet	Model			
	GIGI	G3GI	G4GI	Diet mean
WS	0.908	0.985	0.986	**0.959**
WE3	0.817	0.978	0.929	**0.908**
SB3	0.959	0.990	0.991	**0.959**
Model Mean	**0.916**[a]	**0.984**[b]	**0.972**[b]	
Model s.e.d. (Sig.)	0.0168	(P<0.001)		
Diet s.e.d. (Sig.)	0.0194	(ns)		
M x D s.e.d. (Sig.)	0.0335	(ns)		

Values in the same row with disparate superscripts differ significantly (P<0.05).

Table 2. Rate parameters (λ and K2) and MRT for lambda and K-compartments (LC and KC), time delay (TD) and total tract MRT determined by the G3G1 model of Pond et al. (1988) fitted to faecal excretion data obtained from Yb-marked lucerne silage (WS), enzyme-treated (WE3) and sugar beet pulp substituted (SB3) lucerne silage.

	WS	WE3	SB3	s.e.d.	Sig.
λ	0.337	0.551	0.420	0.1656	ns
LC	9.2	7.0	6.6	1.82	ns
K_2	0.228	0.315	0.128	0.0931	ns
KC	4.8	6.2	8.1	1.16	ns
TD	11.9	15.2	12.3	2.39	ns
MRT	25.9	28.5	27.0	3.15	ns

Discussion

The aims of this experiment were to evaluate the ability of six compartmental models to describe the pattern of faecal excretion in ponies, and select a 'best fit' model for compartmental analysis of the MRT of enzyme-treated or SB-substituted lucerne silage in the different segments of the equid gastrointestinal tract. The G3G1 and G4G1 models of Pond *et al.* (1988) showed the greatest agreement with the faecal excretion data, which concurs with the findings of Moore-Colyer *et al.* (2003) and McLean (2001). Therefore, it appears that time-dependent models have a greater ability to describe the pattern of faecal marker excretion in ponies compared to time-independent models.

There is a paucity of information on the use of mathematical models to compartmentalise the GIT of the horse. Moreover, the limited studies to date have produced inconclusive results (Moore-Colyer *et al.*, 2003), with no clear apportioning of fast and slow compartments. Similarly, no clear distinction could be made between a fast and slow compartment in the current study. Dhanoa *et al.* (1985) also found clear compartmentalisation of the ruminant gut to be problematic despite the model accurately describing eighty sets of ruminant faecal excretion data; therefore, in general, clear biological interpretation of modelled faecal excretion curves appears to be difficult.

The effect of enzyme-treatment or substitutional SB appeared to have no effect on modelled parameters. Currently, no information exists on the effects of enzyme-treatment on the rate of passage of forages in herbivores; although studies have shown an increase in the rate of fibre degradation with the addition of fibrolytic enzymes to forages (Feng *et al.*, 1992, 1996), indicating a possible increase in forage passage rate. Conversely, reduced fibre degradation has been reported for enzyme-treated lucerne fed to ponies (Murray *et al.*, 2007a). Furthermore, little information is available on the MRT of SB substituted forages; in the current study, the MRT of SB3 paralleled that of the control silage; however, it is of note that intakes were restricted and a different scenario may have emerged if these diets were offered *ad libitum*.

Conclusion

Of the six compartmental models evaluated the G3G1 and G4G1 models of Pond *et al.* (1988) showed the greatest agreement with the faecal excretion data; however, it was not possible to draw any definitive conclusions from the compartmentalisation data. Nevertheless, enzyme treatment or SB substitution did not affect the total tract MRT of the lucerne silage, with values ranging from 26 to 28.5 h for all marked diets.

References

Dhanoa, M.S., Siddons, R.C., France, J. and Gale, D.L., 1985. A multicompartmental model to describe marker excretion patterns in ruminant faeces. Br. J. Nutr. 53:663-71.

Feng, P., Hunt, C.W., Julien, W.E., Dickinson, K. and Moen, T., 1992. Effect of enzyme additives in in situ and *in vitro* degradation of mature cool-season grass forage. J. Anim. Sci. 70(1):309 (abstract).

Feng, P., Hunt, C.W., Pritchard, G.T. and Julien, W.E., 1996. Effective of enzyme preparations on in situ and *in vitro* degradation and *in vivo* digestive characteristics of mature cool-season grass forage in beef steers. J. Anim. Sci. 74(6):1349-57.

Grovum, W.L. and Williams, V.J., 1973. Rate of passage of digesta in sheep. 4. Passage of marker through the alimentary tract and the biological relevance of rate constants derived from the changes in concentrations of marker in faeces. Br. J. Nutr. 30:313-29.

McLean, B.M.L., 2001. Methodologies to Determine Digestion of Starch in Ponies [PhD Thesis]: University of Edinburgh, UK.

Moore-Colyer, M.J.S., Morrow, H.J. and Longland, A.C., 2003. Mathematical modelling of digesta passage rate, mean retention time and *in vivo* apparent digestibility of two different chop lengths of hay and big-bale grass silage in ponies. Br. J. Nutr. 90:109-18.

Murray, J.M.D., Longland, A., Davies, D.R., Hastie, P.M., Moore-Colyer, M. and Dunnett, C., 2007a. The effect of enzyme treatment on the nutritive value of lucerne for equids. Livestock Science doi:10.1016/j.livsci.2007.01.156.

Murray, J.M.D., Longland, A., Hastie, P.M., Moore-Colyer, M. and Dunnett, C., 2007b. The nutritive value of sugar beet pulp-substituted lucerne for equids. Anim. Feed Sci. Technol. doi:10.1016/j.anifeedsci.2007.02.013.

Pond, K.R., Ellis, W.C., Matis, J.H., Ferreiro, H.M. and Sutton, J.D., 1988. Compartmental models for estimating attributes of digesta flow in cattle. Br. J. Nutr. 60:571-95.

Siddons, R.C., Paradine, J., Beever, D.E. and Cornell, P.R., 1985. Ytterbium acetate as particulate-phase digesta-flow marker. British Journal of Nutrition. 54:509-19.

Teeter, R.G., Owens, F.N. and Mader, T.L., 1984. Ytterbium Chloride as a Marker for Particulate Matter in the Rumen. J. Anim. Sci. 58(2):465-73.

Growth and glucose/insulin responses of draft cross weanlings fed Total Mixed ration cubes versus hay/concentrate rations

Sarah Ralston[1], Harlan Anderson[2] and Roy Johnson[3]
[1]*Rutgers, the State University of New Jersey, New Brunswick, NJ, USA*
[2] *IdleAcres, Cokato, MN, USA,* [3] *Minnetonka, MN, USA*

Take home message

Draft cross weanlings fed forage based total mixed rations with restricted starch and sugar (NSC) had higher feed efficiency and growth rates than predicted by NRC (1989) and than those fed high NSC concentrates and moderate quality hay. Restriction of NSC will not prevent or resolve all cases of developmental orthopedic disease (DOD) but may help in weanlings with insulin resistance (IR: high plasma insulin responses to glucose challenges). However, not all IR weanlings will develop DOD, regardless of ration.

Introduction

Total mixed rations (TMR), wherein all the nutritional needs of the animals are met in a single feedstuff that is available ad libitum, have been used successfully for decades in growing food animals. However, in the equine industry, the traditional regimen is to provide weanlings concentrates separately and forage in limited amounts, resulting in high starch and sugar (NSC), low fiber meals that cause significant fluctuations in plasma glucose and insulin, which has been hypothesized to adversely impact growth. Increased insulin resistance (IR) has been documented in young horses fed high starch/ sugar feeds (NSC = 20% or higher) and has been correlated with an increased incidence of developmental orthopedic disease (DOD) such as osteochondrosis (OCD), epiphysitis and flexure deformities.

It was hypothesized that rations with low NSC (<20), either as TMR cubes fed free choice or a meal fed concentrate with restricted NSC, would reduce IR and incidence of DOD in weanling horses while sustaining rapid growth rates in draft cross weanlings.

Materials and methods

To test the hypotheses, growth rates, feed efficiency, insulin sensitivity, glucose/insulin responses to the feeds and incidence of DOD in draft cross weanlings were evaluated in a series of three trials conducted in three consecutive years (2004-2006). Each year 12 draft cross weanlings were fed either TMR cubes (Next Generation©, IdleAcres, Cokato, MN) free choice (TMR, n = 6 per year) or hay/concentrate based rations of Nutrena® (Minnetonka, MN) Life Design® Youth® (HS:2004, 2005, n = 6 each year) or Nutrena® (Minnetonka, MN) Safe Choice® (LS:2006, n = 6) to provide 50% of the calories recommended for moderate growth with free choice grass/alfalfa hay for 6 weeks. In 2004 all weanlings were QuarterHorse/Belgian crosses. In 2005 and 2006 three and six of the weanlings, respectively, were more refined Hanoverian/QH/Percheron/TB crosses. Horses were fed in individual stalls overnight and turned out in dry lot paddock 0830-1600h daily. Orts were recorded daily. Horses were weighed and had wither and rump heights recorded weekly. Limbs were visually assessed for epiphysitis and flexure deformities, rated on a scale of 0 (no lesions) to 4 (severe lesions), with scores of 2 or higher considered to be physiologically significant. Radiographs were taken to confirm presence or absence of OCD in hocks and stifles. Insulin sensitivity was assessed with a low dose oral dextrose challenge (LDOD: 0.25 gm dextrose/kg BW) before treatments were initiated and after 6 weeks on treatments (PostTX). Glucose/insulin responses to equicaloric amounts of TMR and the concentrates were measured PostTX in all years. Glucose/insulin data were compared between years by ANOVA for repeated measures factoring effects of treatment, individual and year where appropriate and within year by two tailed Students T-test (Statistixs for Windows, Analytical software). Student t-tests were used to compare feed efficiency (kg gain/Mcal consumed) and average daily gain (kg/day) between treatments within trials. Significance was set at $P < 0.05$.

Results

Nutrient content of the rations differed between and within years (Table 1). All weanlings maintained good general health in all years. No DOD>2 was observed in 2004. In 2005 three horses were IR relative to the others (Insulin responses to LDOD >25 µIU/ml), two of which had DOD>2 (flexure deformities and epiphysitis) before the treatments were initiated. One DOD weanling was placed on TMR (flexure deformity scored 4), the other (epiphysitis 4 and flexure deformity 2) on HC. The weanling fed TMR had a DOD score of 1 within 2 weeks and 0 by the end of the trial, the one fed HC had persistant epiphysitis (scored 4) and flexure deformities (scored 2) throughout the trial. In 2006 two horses were IR, one of which had DOD>2 (Flexure deformity scored 4) that did not change when placed on TMR ration. The other IR weanling had no visual or radiographic lesions. Another weanling fed TMR that was not IR had OCD=4 in the right stifle which was diagnosed at the end of the trial, though it was not evident in the beginning of the study. All others had no significant lesions.

Glucose responses to the PostTX LDOD did not differ between treatments in the first two years but insulin responses tended ($P < 0.1$) to be higher in HS fed horses, suggesting reduced insulin sensitivity though not statistically significant. Glucose/insulin responses to meals of HS were higher ($P < 0.05$) than to TMR or LS. In 2006 TMR fed horses had higher ($P < 0.05$) glucose and insulin responses than those fed LS but both tended ($P < 0.1$) to be lower than in previous years.

Table 1. Nutrient intake-100% DM basis of feed used in the 3 trials.

Year/feed	2004 HG	2004 TMR	2005 HG	2005 TMR	2006 HG	2006 TMR
DE Mcal/kg	2.4	2.2	2.2	2.2	2.6	2.4
% Protein	14.0	18.2	11.0	16.7	14.7	15.6
% ADF	31.8	28.5	29.5	38.8	28.4	35.1
% NSC	20.0	16.5	20.0	13.0	15.4	15.0
% Ca	0.88	1.47	0.73	1.07	1.0	1.18
% Phos	0.45	0.42	0.37	0.35	0.52	0.34
% Mg	0.39	0.32	0.16	0.29	0.31	0.27

Sarah Ralston, Harlan Anderson and Roy Johnson

In all years horses fed TMR had higher (P < 0.05) percent BW average daily gains (%BWADG) and feed efficiency than those on HS (Table 2), though wither and rump heights did not differ (P < 0.05) between treatments.

Discussion

Though the incidence of DOD was low, the data do suggest that horses with IR do not necessarily develop DOD. Genetic predisposition to DOD has been documented in several breeds, which may or may not be correlated with IR. It is of interest to note that three of the DOD horses were refined TB/ Hanoverian/Percheron crosses, the 4th was a very refined ¾ Paint/Belgian cross whereas the non-DOD IR horses were QH/Belgian crosses from bloodlines used in 2004.

The TMR fed horses were consistently more efficient than horses on HG, consuming fewer calories per kg gain in all three years. It is of interest to note that the draft cross weanlings, regardless of dietary treatment, voluntarily consumed < 90% of NRC (1989) recommended calories for moderate growth (> 600kg mature weight) but sustained growth rates 90-143% of the 0.8kg/day predicted. In 2005 the quality of the hay was only moderate and the differences between treatment groups were greater than in 2004 and 2006. It is interesting that despite the apparent similarities in intakes in 2006 that the TMR fed horses

Table 2. Growth of weanlings (n=6/treatment/yr) fed either HG or TMR for 5 weeks.

Year/Feed	2004 HG	2004 TMR	2005 HG	2005 TMR	2006 HG	2006 TMR
%BW intake	2.8±0.03[a]	2.9±0.04[a]	2.7±0.04[a]	2.9±0.06[a**]	2.7±0.05[b]	2.8±0.05[b*]
%BWADG	0.31±0.01[a]	0.37±0.03[a]	0.29±0.01[a,b]	0.36±0.03[a,b**]	0.27±0.03[b]	0.38±0.02[b]
Efficiency	0.11±0.01[a]	0.16±0.01[a**]	0.07±0.01[a]	0.12±0.02[a**]	0.09±0.01[a]	0.15±0.01[a*]
%NRC req	85.6±1.0[a]	71.3±1.2[a**]	88.6±0.9[a]	71.8±1.3[a**]	84.3±1.4[a]	78.4±1.1[a**]

%BWADG=(ADG (kg/day)/BW(kg))*100, Efficiency=ADG/Mcal consumed, NRCreq=% of NRC (1989) recommended daily Mcal intake for 0.8Kg gain/day.
[a,b] Means with different letters differ between years P<0.05.
*Values differ between treatments within the year P<0.05.
**Values differ between treatments within the year P<0.01.

236 | *Applied equine nutrition and training*

were still more efficient than the horses on LS. It is interesting to note that%BWADG was lower and feed efficiency tended ($P<0.01$) to be lower in the years that more horses with a lower percent of draft blood were used (2005, 2006).

Both rations in 2005 and the TMR in 2006 had marginally lower phosphorus (0.34-0.37% DM) than recommended for growth (0.40% DM) but the apparent deficits were not clearly associated with the presence of DOD or reduced growth rates. Current recommendations may be in excess of actual needs.

Conclusions

Feeding TMR cubes formulated for growth free choice is an efficient alternative to traditional hay/high NSC concentrate rations under the conditions of these trials. The 1989 NRC caloric recommendations for growth may be in excess of the needs of draft cross weanlings. The responses seen in 2005 suggest that placing IR horses with flexure deformities and/or epiphysitis on low NSC rations may help to resolve the problems, but further research is necessary. Restriction of NSC will not prevent or resolve all cases of DOD and not all IR weanlings will develop DOD if fed rations formulated for growth.

Energy intakes of three different equine populations in comparison to NRC recommendations

K. Wilkinson, J. Holley, C. Preston and R.J. Williams
Hartpury College, University of the West of England, Gloucester, GL19 3BE, United Kingdom

Take home message

Actual energy intakes of racehorses most closely reflected calculated NRC values, while other groups (riding school and livery horses) were fed diets which resulted in energy intakes that were greater than NRC values. In some groups of horse this can lead to an increase in bodyweight and suggests the need for increased education on feeding practice within the horse owning public. Racehorses were not fed according to bodyweight and therefore do not meet NRC feeding recommendations.

Introduction

The most commonly used reference energy requirements for horses are based on the NRC recommendations (1989). The guidelines use the DE system to calculate energy intake as a proportion of bodyweight to estimate minimum energy requirements to maintain health. The NRC guidelines do make allowances for activity, heavy horses (>600Kg) (Potter *et al.*, 1987) however do not take into consideration environmental effects or individual metabolic variation. More recent research has added a variation factor for horses living in cold conditions (Cymbaluk and Christison, 1990).

Adherence to NRC guidelines in different horse populations has not been thoroughly investigated. Management strategies vary depending on workload, environment and intended use, ie. pleasure or competition. The aims of the study were to (1) compare the actual energy intake of horses in three populations to the NRC guidelines, and (2) to establish if any dissociation from the NRC value resulted in a change in bodyweight.

Materials and methods

Eighty-one horses from two yards were used for the study (Yard 1: 30 National Hunt racehorses (RH), Yard 2: 26 Riding School horses (RS), 25 private liveried horses (DIY) between October and December 2005. Total daily energy was calculated on 5 days over a 1 month period using manufacturers DE values and weight of feed fed. Horses were fed as normal (combination of concentrates and forage) with no dietary change during the study; concentrate to forage ratio was calculated. Hay intake was measured morning and evening on five separate occasions and a mean calculated for each horse. Any left over hay was weighed and subtracted from the total daily intake. The type and make of each feed given to the horses was also recorded. Two samples of hay were collected from each yard and sent for independent laboratory analysis for DE values. Horses did not have access to turnout and had been managed in this way for at least two months prior to the study; water was available *ad libitum*. The bodyweights of all horses were measured at weekly intervals throughout the collection period. Average workload of each horse was estimated (light (RS), medium (DIY), hard (RH) during the study period by assessing the daily work done according to the NRC guidelines (1989) adapted by Frape (2004). Data was used to calculate the NRC and actual energy intakes of the study populations. Wilcoxon signed rank test and Pearsons correlation were used to compare NRC and actual energy values, and NRC-actual energy and bodyweight respectively. Forage and concentrate ratios were analysed using a 1 way ANOVA and post hoc Tukey test.

Results

There was no difference in the actual energy intake of the RH group (146.1 ± 4.2DE MJ / day) and the NRC value (147.9 ± 7.0 MJ DE/day), however, there was a trend for the actual intake to be lower than the NRC minimum recommended value ($P > 0.05$). Both the RS (121.9 ± 23.9 MJ DE/day) and DIY (122.2 ± 28.9 MJ DE/day) groups had a significantly greater energy intake than the calculated NRC value (94.5 ± 14.2; 96.3 ± 12.6 MJ DE/day) ($P < 0.01$). There was also a greater range in the energy intakes of the DIY (73.86 – 186.65 MJ DE

/day) and RS (88.78 – 170.86 MJ DE/day) groups in comparison to the RH group (135.0-153.4 MJ DE/day) although this corresponded to a greater variety of work scores and bodyweights in comparison to the racehorse population (P < 0.05) (Figure 1 and 2).

Bodyweight and actual energy intake was positively correlated in the RS (P < 0.01) and DIY (P < 0.05) groups. No significant correlation between bodyweight and actual energy intake was found in the RH group (P > 0.05).

The change in bodyweight over the study period and difference between NRC and actual energy values (NRC-actual energy intake) was negatively correlated in the DIY group (P < 0.05) indicating that the positive energy balance resulted in an increase in bodyweight. There was no significant correlation between the change in bodyweight seen

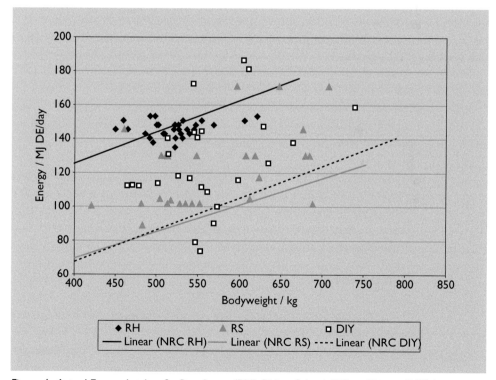

Figure 1. Actual Energy Intakes for Racehorse (RH), Riding School (RS) and Livery (DIY) horses over a range of bodyweights compared to linear NRC guideline values.

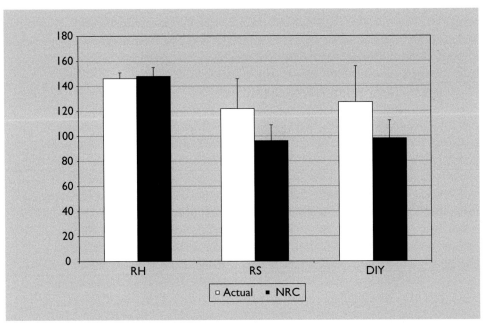

Figure 2. Mean actual and NRC energy intakes (MJ DE/day) (± s.d.) in racehorse (RH), riding school (RS) and livery (DIY) populations.

and the difference between NRC and actual energy values (NRC-actual energy intake) in the RH and RS groups (P > 0.05), however, a negative trend was seen in the RH group.

Significant differences in the forage to concentrate ratios between RH (34:66%), RS (75:25%) and DIY (88:12%) groups were seen (p < 0.01). The difference between the NRC recommendation and the actual forage to concentrate ratio was shown to be significant in the DIY group only (P < 0.05).

Discussion

As a group, racehorses were fed most closely to NRC calculated values, however, there was a trend for these horses to be fed below the recommended minimum energy intakes. Despite the range of bodyweights found within this group (448-622 kg), all horses were fed exactly the same concentrate ration, which may mean that some of the heavier horses were being underfed, with respect to energy intake.

There was no correlation between the difference between the NRC and actual energy values (NRC-actual energy intake) and the change in bodyweight in either the RH or RS groups which suggests that the other factors associated with energy requirements and utilisation affect the accuracy of the NRC guidelines.

Both the RS and DIY groups were fed a diet with a greater energy content than recommended by the NRC ($P < 0.01$); RS data in the present study were similar to previous studies on a riding school population (Jansen *et al.*, 2001) despite the use of a different energy system (NE versus DE). The DIY group received 30% greater energy than the recommended NRC guidelines which suggests that they were being overfed and this was reflected in the significant rise in bodyweight seen ($P < 0.05$). No changes were seen in the bodyweights of the RS group, even though the energy intake exceeded the minimum NRC guidelines. This suggests that the individualised feeding approach used in the RS group was able to balance energy intake with workload and type of horse.

Forage to concentrate ratios revealed that RH group were fed at NRC guidelines, receiving 34% forage (NRC 35%), however this is well below the minimum forage guidelines suggested for optimum hind gut health, passage rates and behaviour. The average forage intake for the RS and DIY groups were 75% and 88% respectively. This suggests that both populations are on high forage diets which is beneficial to gastrointestinal health and for mimicking natural foraging time budget behaviour. The forage intake in the DIY group is higher than recommended for horses in medium work, and therefore may have contributed to the weight gain seen over the study period. Although the positive energy balance was due to increased levels of forage fed, it is preferable to feed too much forage than too much starch (in the form of concentrates) in the diet of stabled horses with limited/no turnout.

Conclusions

The actual energy intake of horses in the DIY group deviated most from the NRC recommendations and this was reflected in an increased bodyweight. Further investigation of feeding practices within this group is recommended. This study gives valuable information for veterinary surgeons and nutritionists involved in the management of different horse populations and suggests that further investigation into feeding practices and the use of bodyweight in ration formulation is warranted.

References

Cymbaluk, N.F. and Christison, G.L., 1990. Environmental effects on thermoregulation and nutrition of horses. Veterinary clinics of North America: Equine Practice 6: 355-372.

Frape, D., 2004. Equine Nutrition and Feeding. Blackwell: Oxford.

Jansen, W.L., van Alphen, M., Berghout, M., Everts, H. and Beynen, A.C., 2001. An approach to assessment of the efficiency of dietary energy utilisation by horses and ponies kept at riding schools. The Veterinary Quarterly 23(4): 195-198.

NRC, 1989. Nutritional requirements of horses. 5th National Academy Press, Washington, US.

Potter, G.D., Evans, J.W, Webb, G.W and Webb, S.P., 1987. Digestible energy requirements of Belgian and PErcheron Horses. In: Proceedings of the 10th Equine Nutrition and Physiology Symposium, Fort Collins, Colorado. p133-138.

Printed in the United States
by Baker & Taylor Publisher Services